Springer Series in Wireless Technology

Series editor

Ramjee Prasad, Aalborg, Denmark

Springer Series in Wireless Technology explores the cutting edge of mobile telecommunications technologies. The series includes monographs and review volumes as well as textbooks for advanced and graduate students. The books in the series will be of interest also to professionals working in the telecommunications and computing industries. Under the guidance of its editor, Professor Ramjee Prasad of the Center for TeleInFrastruktur (CTIF), Aalborg University, the series will publish books of the highest quality and topical interest in wireless communications.

More information about this series at http://www.springer.com/series/14020

Ramjee Prasad · Sudhir Dixit
Editors

Wireless World in 2050 and Beyond: A Window into the Future!

Editors
Ramjee Prasad
Center for TeleInFrastruktur
Aalborg University
Aalborg
Denmark

Sudhir Dixit
Skydoot, Inc
Woodside, CA
USA

ISSN 2365-4139 ISSN 2365-4147 (electronic)
Springer Series in Wireless Technology
ISBN 978-3-319-82508-3 ISBN 978-3-319-42141-4 (eBook)
DOI 10.1007/978-3-319-42141-4

Softcover reprint of the hardcover 1st edition 2016

Printed on acid-free paper

This Springer imprint is published by Springer Nature
The registered company is Springer International Publishing AG Switzerland

सहजं कर्म कौन्तेय सदोषमपि न त्यजेत्
सर्वारम्भा हि दोषेण धूमेनाग्निरिवावृताः

saha-jam karma kaunteya
sa-dosam api na tyajet
sarvarambha hi dosena
dhumenagnir ivavrtah

Bhagavad Gita 18.48

Every endeavor is covered by some sort of fault, just as fire is covered by smoke. Therefore one should not give up the work which is born of his nature, O son of Kuntī, even if such work is full of fault.

To all the past, present, and future Ph.D. researchers

Preface

This book is an outcome of a day-long seminar on *wireless and human bond communications beyond the year 2050* held on June 22, 2015 at the Center for Teleinfrastruktur, Department of Electronic Systems, Aalborg University, Denmark, to commemorate the supervision of 100 Ph.D. students by Ramjee Prasad. During the event, 18 presentations were made by the former Ph.D. students and the invited speakers who have attained prominence in their respective fields of endeavors around the world. The talks covered a broad range of topics related to wireless communications. The core theme of the seminar was to fast-forward the future to 2050 and speculate the innovations yet to come, how society would adopt to changes, and what types of strategies and business models would likely be in play in about 35 years' time from now. In short, the seminar was an exercise to create a time capsule for the future generations to look back and determine how correct or wrong the predictions were. The event was such a success that it was decided that the speakers write up position papers on the ideas that they presented. These were then adapted into book chapters to be published collectively in a book.

In addition to discussing overall transformation of the wireless systems of the future from a historical perspective, some authors go into details of likely technological innovations to come and how they would impact certain industries, such as healthcare, automobile, IoT/M2M, entertainment, and the society overall. The past decade has seen many changes in the business models for revenue generation. This trend is going to pick up even more momentum in the future. Therefore, we have chapters included on multi-business models and business strategies to meet the challenges for the 2050 time-frame.

The book is intended for casual readers not necessarily familiar with wireless technologies. Therefore, the content is more descriptive and qualitative than theoretical in style of writing.

We (including the chapter authors) have made every effort to be as accurate as possible, but some errors are inevitable. We encourage the readers to let us know of any errors, which we will correct in future editions of the book.

Aalborg, Denmark
Woodside, USA
June 2016

Ramjee Prasad
Sudhir Dixit

Acknowledgments

We thank the contributors of this book for their time and effort to make this book possible in a short period. They readily agreed to write chapters based on their presentations at a day-long seminar on June 22, 2015, organized at the Aalborg University, to celebrate the successful supervision of the 100 Ph.D. graduates by Ramjee Prasad. The seminar focused on the innovations yet to come in the world of wireless in the beyond 2050 period and the new and emerging field of human bond communication.

Contents

About the Editors

Dr. Ramjee Prasad is the Founder President of the CTIF Global Capsule (CGC). He has been Founding Director of Center for Teleinfrastruktur (CTIF) since 2004. He is also the Founder Chairman of the Global ICT Standardisation Forum for India, established in 2009. GISFI has the purpose of increasing of the collaboration between European, Indian, Japanese, North American, and other worldwide standardization activities in the area of Information and Communication Technology (ICT) and related application areas.

He was the Founder Chairman of the HERMES Partnership—a network of leading independent European research centers established in 1997, of which he is now the Honorary Chair. He is a Fellow of IEEE (USA), IETE (India), IET (UK), Wireless World Research Forum (WWRF) and a member of the Netherlands Electronics and Radio Society (NERG), and the Danish Engineering Society (IDA).

He has received Ridderkorset af Dannebrogordenen (Knight of the Dannebrog) in 2010 from the Danish Queen for the internationalization of top-class telecommunication research and education. He has been honored by the University of Rome "Tor Vergata," Italy as Distinguished Professor of the Department of Clinical Sciences and Translational Medicine on March 15, 2016.

He has received several international awards as follows: IEEE Communications Society Wireless Communications Technical Committee Recognition Award in 2003 for making contribution in the field of "Personal, Wireless and Mobile Systems and Networks"; Telenor's Research Award in 2005 for impressive merits, both academic and organizational within the field of wireless and personal communication; 2014 IEEE AESS Outstanding Organizational Leadership Award for "Organizational Leadership in developing and globalizing the CTIF (Center for TeleInFrastruktur) Research Network," and so on.

He is the Founder Editor-in-Chief of the Springer International Journal on Wireless Personal Communications. He is a member of the editorial board of other renowned international journals including those of River Publishers. Ramjee Prasad is Founder Co-Chair of the steering committees of many renowned annual international conferences, e.g., Wireless Personal Multimedia Communications Symposium (WPMC); Wireless VITAE and Global Wireless Summit (GWS).

He has published more than 30 books, 1000 plus journal and conference publications, holds more than 15 patents, and has supervised over 100 Ph.D. graduates and larger number of masters (over 250). Several of his students are today worldwide telecommunication leaders themselves.

Dr. Sudhir Dixit is the CEO and a Co-Founder of Skydoot, Inc., a start-up in the content sharing and collaboration space. He is also a Fellow and Evangelist of Basic Internet at the Basic Internet Foundation in Norway. From December 2013 to April 2015, he was Distinguished Chief Technologist and CTO of the Communications and Media Services for the Americas Region of Hewlett-Packard Enterprise Services in Palo Alto, CA, and prior to this he was the Director of Hewlett-Packard Labs India from September 2009. From June 2009 to August 2009, he was Director at HP Labs in Palo Alto. Prior to joining HP Labs Palo Alto, Dixit held a joint appointment with the Centre for Internet Excellence (CiE) and the Centre for Wireless Communications (CWC) at the University of Oulu, Finland. From 1996 to 2008, he held various positions with leading companies, such as with BlackBerry as Senior Director (2008), with Nokia and Nokia Networks in the United States as Senior Research Manager, Nokia Research Fellow, Head of Nokia Research Center (Boston), and Head of Network Technology (USA) (1996–2008). From 1987 to 1996, he was at NYNEX Science and Technology and GTE Laboratories (both now Verizon Communications) as Staff Director and Principal Research Scientist.

Sudhir Dixit has 21 patents granted by the US PTO and has published over 200 papers and edited, co-edited, or authored six books (Wi-Fi, WiMAX and LTE Multi-hop Mesh Networks by Wiley (2013), Globalization of Mobile and Wireless Communications by Springer (2011), Technologies for Home Networking by Wiley (2008), Content Networking in the Mobile Internet by Wiley (2004), IP over WDM by Wiley (2003), Wireless IP and Building the Mobile Internet by Artech House (2002)). He is presently on the editorial boards of IEEE Spectrum Magazine, Cambridge University Press Wireless Series, and Springer's Wireless Personal Communications Journal and Central European Journal of Computer Science (CEJS). He is Chairman of the Vision Committee and Vice Chair of the Americas region of the Wireless World Research Forum (WWRF). He also chairs the IEEE ComSoc Sub-Technical Committee on Fiber and Wireless Convergence.

From 2010 to 2012, he was Adjunct Professor of Computer Science at the University of California, Davis, and, since 2010, he has been Docent of Broadband Mobile Communications for Emerging Economies at the University of Oulu, Finland. A Fellow of the IEEE, IET, and IETE, Dixit received a Ph.D. degree in Electronic Science and Telecommunications from the University of Strathclyde, Glasgow, U.K. and an M.B.A. from the Florida Institute of Technology, Melbourne, Florida. He received his M.E. degree in Electronics Engineering from Birla Institute of Technology and Science, Pilani, India, and B.E. degree from Maulana Azad National Institute of Technology, Bhopal, India.

Abbreviations

3D	Three Dimensional
3GPP	3rd Generation Partnership Project
ABP	Arterial Blood Pressure
ADC	Analog-to-Digital Converter
ADPCM	Adaptive Differential Pulse Code Modulation
AFCRN	Absolute Radio Frequency Number
AI	Artificial Intelligence
AN	Access Node
ANN	Artificial Neural Network
AON	Active Optical Network
AP	Access Point
ARM	Association Rule Mining
BBU	Base-band Units
BD	Big Data
BLE	Bluetooth Low Energy
BM	Business Model
BMES	Business Model Ecosystems
BPSK	Binary Phase-Shift Keying
BS	Base Stations
CAD	Computer-Aided Design
CAPEX	Capital Expenditure
CD	Compact Disk
CDM	Code-Division Multiplexing
CDSS	Clinical Decision Support Systems
CELP	Code Excited Linear Predictive
CN	Core Network
CNSS	Communications Navigation Sensing and Services
COMSOC	Communications Society (IEEE)
COTS	Commercially off the shelve
CPS	Cyber-Physical Systems

CR	Cognitive Radio
C-RAN	Cloud Radio Access Networks
DB	Distributed Beamforming
DCR	Distant cognitive radio
E_b/N_0	Bit Energy-to-Noise Power Spectral Density Ratio
EC	European Commission
ECG	Electrocardiogram
EDGE	Enhanced Data rates for GSM Evolution
EEG	Electroencephalogram
EMR	Electronic Medical Records
ESPAR	Electronic Steerable Parasitic Antenna Array
FiWi	Fiber-Wireless
FPLMTS	Future Public Land Mobile Telecommunication Systems
FTTx	Fiber to the "x"
GDP	Gross Domestic Product
GPRS	General Packet Radio Service
GPS	Global Positioning System
GPU	Graphics Processing Unit
GSM	Global System for Mobile Communications
HBC	Human Bond Communication
HetNets	Heterogeneous Networks
HSDPA	High-Speed Downlink Packet Access
HSPA	High-Speed Packet Access
HVDC	High-Voltage Direct Current
ICD	International Classification of Diseases
ICT	Information and Communication Technologies
ICU	Intensive Care Unit
IEEE	Institute of Electrical and Electronics Engineers
IMT-2000	International Mobile Telecommunications-2000
IoE	Internet of Everything
IoS	Internet of Services
IoT	Internet of Things
IP	Intellectual Property
IS-54 (136)	Industry Standard-54 (136) (in USA)
ISM	Industrial Scientific Medical
ITS	Intelligent Transportation Systems
ITU	International Telecommunication Union
ITU-T	ITU Telecommunication Standardization Sector
KPI	Key Performance Indicator
LINP	Logically Isolated Network Partitions
LMSC	LAN/MAN Standardization Committee
LNA	Low-Noise Amplifier
LoRa	**Lo**w power long **Ra**nge
LP	Long Play

LTE	Long-Term Evolution
LTE-A	Long-Term Evolution Advanced
M2M	Machine 2 Machine
MIC	Mutual Information Coefficient
MIMIC	Multiparameter Intelligent Monitoring in Intensive Care
MIMO	Multiple Input–Multiple Output
MIT	Massachusetts Institute of Technology
MME	Mobility Management Entity
MtM	Machine-to-Machine
NFV	Network Function Virtualization
NGMN	Next-Generation Mobile Network
Ng-PON	Next-Generation Passive Optical Networks
NLP	Natural Language Processing
OAM&P	Operation, Administration, Maintenance, and Provisioning
OFDM	Orthogonal Frequency-Division Multiplexing
ONT	Optical Network Terminal
OPEX	Operational Expenditure
OSI	Open Systems Interconnection
OTT	Over the Top
PA	Power Amplifier
PAPR	Peak-to-Average Power Ratio
PCA	Principal Component Analysis
PGW	Packet Gateway
PHS	Personal Handy Phone System
PLL	Phase Lock Loop
PON	Passive Optical Network
PPG	Photoplethysmogram
PtP	Point-to-Point
PU	Primary User
QAM	Quadrature Axis Modulation
QoS	Quality of Service
R&D	Research & Development
RAN	Radio Access Network
RBM	Restricted Boltzmann Machine
RC	Radio Committee
RDBMS	Relational DataBase Management Systems
RF	Radio Frequency
RRU	Remote Radio Unit
SCM	Sub-Carrier Multiplexing
SDN	Software-Defined Networking
SDR	Software-Defined Radio
SGW	Serving Gateway
SISO	Single Input Single Output
SMNAT	Smart Mobile Network Access Topology
SON	Self-Organizing Networks

SpO_2	Saturation of Peripheral Oxygen
SSC	Satellite and Space Communications Committee
SSD	Solid-State Disk
STEAM	Science, Technology, Engineering, Arts, and Mathematics
SVD	Singular Value Decomposition
SVM	Support Vector Machine
TCUP	TCS Connected Universe Platform
TDM	Time Division Multiplexing
TV	Television
UAV	Unmanned Aerial Vehicle
UICC	Universal Integrated Circuit Card
UMLS	Unified Medical Language System
UMTS	Universal Mobile Telecommunications System
USB	Universal Serial Bus
UVI	Unified Virtualized Interface
UVN	Unified Virtual Network
V2I	Vehicle-to-Infrastructure
V2V	Vehicle-to-Vehicle
VAS	Value-Added Services
VBSP	Virtual Base Station Pool
VCELP	Vector CELP
VLSI	Very Large Scale Integration
VoLTE	Voice over LTE
W-CDMA	Wideband-Code Division Multiple Access
WDM	Wavelength Division Multiplexing
WLAN	Wireless Local Area Networks
WPAN	Wireless Personal Area Networks
WSN	Wireless Sensor Network

Chapter 1
Introduction

Sudhir Dixit and Ramjee Prasad

Broadband mobile communication has grown by leaps and bounds over the past 15 years that is unprecedented in the history of telecommunication. With the unleashing of open interfaces, open development platforms, open internet and web technologies, the developers and the users are playing an important role in how and what services and applications are created and consumed at breadth taking pace. Adjacent technologies of big data, virtualization, analytics, and multimedia are only going to intensify transformation towards the digital world over the next 35 years or so. Consequently, it is hard to imagine how the following aspects of the entire wireless eco-system are going to change: (i) network infrastructure, (ii) network services, (iii) end-user applications (both in the consumer and business domains), (iv) spectrum management, (v) context- and location-aware personalization, (vi) user interfaces, strategies for planning and deployment, and (vii) business models.

This book is an attempt to predict how the technologies of wireless networking and the networks themselves are likely to evolve over time (or be replaced completely by something totally different) in about 35 years' time. Whenever appropriate the authors have looked at the historical trends to derive cues for the future. Figure 1.1 illustrates some of the key technologies that will undergo significant innovation and will have a profound impact on wireless networks and the accompanying services in the year 2050 time-frame. These are discussed in the various chapters of the book. In Chap. 14, Epilogue, we highlight the key anticipated innovations in these areas.

S. Dixit (✉)
Skydoot, Inc., Woodside, USA
e-mail: Sudhir.dixit@ieee.org

R. Prasad
Aalborg University, Aalborg, Denmark
e-mail: prasad@es.aau.dk

R. Prasad and S. Dixit (eds.), *Wireless World in 2050 and Beyond: A Window into the Future!*, Springer Series in Wireless Technology,
DOI 10.1007/978-3-319-42141-4_1

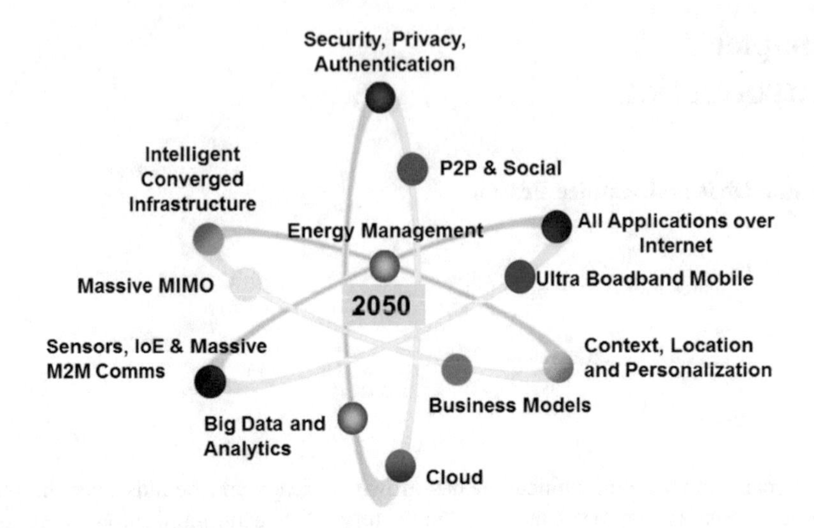

Fig. 1.1 Key technology enablers for wireless communications and systems in the year 2050

1.1 Brief Overview of the Book

The book has been organized to cover a wide range of topics related to wireless communications that are likely to undergo significant innovation and will drive the wireless networks and services in the year 2050 time-frame. Information technology and telecommunications will converge into one and the only reminiscent of the past will be the wires or the radio access to interconnect the computing devices (which would also be the networking devices as we call them today).

In this chapter is an introduction of the book and describes the goals of the book and the key areas of major innovation likely to occur between now and the year 2050. The authors of the chapters dwell into more details in these selected areas.

Chapter 2 describes the major technology advancements in the last 50 to 100 years and predicts the advancements by year 2050. The author also discusses the continuation of the present socio-economic environment and its possible evolution into something different that might impact the innovation environment. He also discusses the societal-government balance that may either result into a closed environment (dystopian world) or an open environment (utopian world), which would likely affect the evolution (or development of new) technologies and services and solutions available to users.

Chapter 3 describes that in the future there will be massive numbers of things and machines connected along with the humans and these would require continuous increase in the data rates of the access networks. The ways in which these would be connected will put new requirements on next generation networks. The author presents an overview of the current access network architectures and proposes a next generation access network topology. Fiber and wireless will be integrated in access as never before. After considering and analyzing all the expected issues the author proposes the idea of a future core and access network that will evolve into a Unified Virtual Network, which will have unique and unified access infrastructure, based on Unified Virtual Access Points for wired and wireless access.

In Chap. 4, the author looks back at the wireless communications R&D history and speculates what will happen 35 years from now in 2050. He makes the case that instead of focusing exclusively on technology development in vacuum, it should be driven by applications and services and this should define what approaches the companies should follow.

In Chap. 5, the author proposes that smart physical layer addressing and mobility management schemes, such as Smart Mobile Network Access Topology (SMNAT), will enable device-to-device communication in such a way that a physical mobile network as we know it today will not be needed, heralding an era of free and unlimited connectivity without networks. The addressing and mobility management will be made possible by the physical layer. This would, however, require non-intelligent wireless access points for boosting pilot signals throughout the coverage area. Some simple network elements would be required for the purposes of authentication, service authorisation, and usage metering for billing.

Chapter 6 is about wireless sensor networks (WSNs). The author describes its state-of-the-art and how WSNs are being deployed and utilized today in a range of applications. He then presents the various enabling technologies for the advancement and development of future WSNs. Some examples of the technologies discussed are distributed beam-forming, cognitive radio, joint sensing and communication with OFDM, physical layer security as well as wavelet technology for context-aware WSNs. Finally, the author makes the case that flexible software structures will be needed to reconfigure the signal setting and implementation to make them adaptable such that they transform into intelligent WSNs.

Chapter 7 describes that in the future everything will be connected with the cloud over high speed access networks, and the cell sites, server racks, antennas and radio equipment, and everything else would need to become small. The resulting systems would need to be efficient not only in size but in spectrum and energy uses, mainly due to lack of resources. To meet these challenges, the author proposes higher order beam space MIMO with joint antenna techniques to reach similar performance as in conventional multiple-input multiple-output (MIMO) but at half the energy consumption with much smaller form factor. This has been demonstrated with 16-quadrature amplitude modulation (QAM) beam space MIMO.

Chapter 8 is about how the megatrend of digitalization fueled by the proliferation of the IoT will completely change the way we live, play, work, pay, travel and drive. It will have a phenomenal impact on every industry vertical. The author uses the example of the automotive industry to explain what it means to all the actors in this domain. The IoT is rapidly transforming the automotive industry from more automation and sofwarization to connected car to full autonomous car, enabling new consumption and monetization models. Some examples of these monetization models are car sharing, pay-as-you-drive, etc. The automobile will need to be always connected to 5G and beyond systems to exploit the data and intelligence available from the Internet-based services. From the mobile broadband connectivity perspective, there are a number of challenges that will need to be addressed for an automobile to become fully autonomous and support all the autonomous driving use cases that are discussed in the chapter.

In Chap. 9, the authors describe how future healthcare systems are evolving as a result of current research and development in Internet of Things (IoT), connected devices and sensors, signal processing, machine learning and artificial intelligence. Collecting patient data through sensors and collectively analysing them over a long period for a large set of patient population has the potential to discover new disease diagnostic and treatment protocols. Leveraging sensor data, analytics, AI and machine learning offers a great opportunity to change the healthcare approach from being illness-driven to wellness-driven. The authors predict that these technologies and such new approaches would be prevalent in the year 2050 time-frame.

Chapter 10 is about the challenges and opportunities ahead with the convergence of IoT and big data and analytics in the various industry verticals. In addition to energy-efficient miniaturized sensors, wireless connectivity and cellular networks have made all this possible (to deploy sensors anywhere). However, lack of standards and lack of motivation by the various players, who need to cooperate, is going to be barriers toward rapid and widespread deployment. Finally, the author presents some major innovations likely to happen in the decade of 2050, and presents business development ideas and steps that an innovator and entrepreneur can take now to start and accelerate a successful career in big data and IoT that could easily last beyond the 2050s.

In Chap. 11 the authors proclaim that one cyber year is equivalent to 7 human years, meaning that what we predict in human year 2050 will actually be happening in the year 2020 simply because the pace of technology development is fast and furious. This pace puts security at a significant stress, requiring the industry to re-think the overall security model. The authors present the cyber security perception and approaches in the past, present, and the future. Thereafter they present their vision of the society in 2050 and its implications on cyber security, followed by solutions to security and privacy. The authors conclude that the tussle between security, privacy and law enforcement will continue to prevail well beyond the year 2050.

Chapter 12 describes the history of telecommunications, its major milestones, strategies used to get where we are today which is a new digital era. The term telecommunications has evolved to be called communications driven by the continuous and disruptive innovations in the Internet. Although 2050 might seem a long time away, it is the perfect time to set in motion the strategies to be successful in the 2050 time-frame. Therefore, the author proposes 10 strategic steps that should be taken today to set the fundamentals right to be a market winner in the long-term!

In Chap. 13, the author introduces the concept and theory of persuasive multi-business models and applies them to the upcoming 5G mobile networks. He explains how businesses can use these models and what they really can do with them to be successful. Finally, the author predicts that persuasive business models will be in widespread use in 2050 and beyond, and would eventually be based on a common understanding and vision of the secure world, resulting in secure persuasive business models.

In Chap. 14, the editors summarize the book by segmenting the technologies into specific domains, as illustrated in the atom diagram of Fig. 1.1. They highlight the challenges and need for research in key areas to achieve the vision of networking and communications in the year 2050 and beyond. They also touch upon the evolution of business models and societal impacts between now and then.

Chapter 2
Technology Advancements in 2050 and How the World Will Look Like

Alexandru Vulpe

Abstract The last 100 years have shown technological progress at a tremendous pace. Simultaneously, the social, economic and political changes have also been influenced by the technological progress but at a slower pace. The present work makes the daring exercise of envisioning how the technological landscape will look like in the year 2050, not from a purely technical viewpoint, but also taking into account social, economic and political aspects. We come to the conclusion that there will be either a closed world, where access to technology is strictly controlled and innovation is possible only under a controlled environment or an open world where there is unrestricted access to technology and innovation continues to thrive, by being built on open systems.

If we take into account technology advancement in the last 50–100 years, we can imagine, unless some major setback occurs, that it will continue ever more rapidly. It is likely that discreet and unobtrusive technological advances, devices and information overlays will change how we live in significant ways. But what will all this new technology mean? Will advances in technology make us more empowered, motivated and active, rather than passive consumers of infotainment? Will threats endanger much of the openness that we now enjoy online?

Before we envision how 2050 will be like, we must start with some assumptions. The first one is that the current capitalist system remains until 2050, meaning there will still be a free market in the majority of countries, and people and enterprises will continue to innovate. The second one is that global warming impacts are not catastrophic and there will be no second ice age. Next, we assume that global economy will not be in a permanent recession and companies continue to innovate and grow. Lastly, we must assume that nanotechnology becomes effective, enabling tiny devices to be embedded inside the human body.

A. Vulpe (✉)
University Politehnica of Bucharest, Bucharest, Romania
e-mail: alex.vulpe@radio.pub.ro

R. Prasad and S. Dixit (eds.), *Wireless World in 2050 and Beyond: A Window into the Future!*, Springer Series in Wireless Technology,
DOI 10.1007/978-3-319-42141-4_2

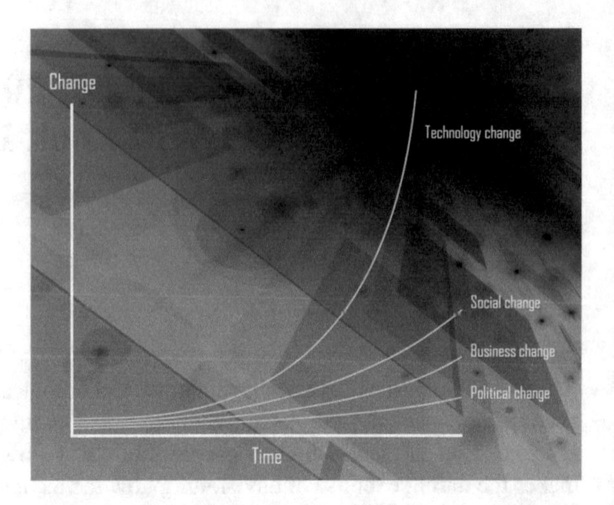

Fig. 2.1 Changes over time [2]

As highlighted in Fig. 2.1, the most dramatic change over time is the change in technology. We have already witnessed this in the last 50–100 years, and there is no reason to think it will stop. On smaller scale, in part triggered by the technological changes, we can say there are social changes, business changes and political changes. If we think about the influence of technology, social changes occur because of the possibility for people to (almost) instantly connect one with another, regardless of the geographical distance between them. Business changes are triggered, for example, by current paradigms of ubiquitous computing and rise of interconnected machines (cyber-physical systems). Lastly, political change is influenced by technological change in the sense that new technologies have to be regulated, standardized and they also influence the way politics is being made.

Our proposed exercise of envisioning how the world will look like in 2050 will, as detailed in the next sections lead to two different scenarios. One scenario is that there will be a closed, dystopian world, where people only know what stakeholders (governments, large corporations) want them to know and are served specifically tailored content, transforming them into informational "couch-potatoes". The other, more optimistic, scenario is a utopian world, where open systems continue to exist and serve as means for people to become more actively involved in the problems of society, empowering different marginalized groups and societies.

The chapter is organized as follows. Section 2.1 presents an overview of how technology and the Internet will look like in 2050. Section 2.2 present how this will shape Industry and economy in 2050, while Sect. 2.3 envisions what happens when we lose access to technology. Section 2.4 discusses open and closed systems and, based on this discussion Sect. 2.5 outlines two possible scenarios of future worlds. Finally, Sect. 2.6 draws the conclusions.

2.1 Technology and Internet in 2050

2.1.1 Introduction

In order to make the exercise of predicting technology advancement in the next 40 years or so, we must look at how much technology has advanced in the last 40 years. In other words, we must look at the past in order to predict the future.

One of the most popular indications of technology advancement is linked with the so-called Moore's law [1]. The law represents an "observation that the number of transistors in a dense integrated circuit doubles approximately every 2 years". It is most times quoted as a "doubling of chip performance every 18 months" (as a combination of the increase of the number of transistors in a chip and the increase in their speed). Figure 2.2 shows a potential evolution of transistor size up until 2050.

It is widely considered that the asymptotic limit of the size of a transistor is 10 times the molecular size, or 10 nm. We can see from Fig. 2.2 that this limit is approximately reached in 2030. Will this hinder technology advancement? Most likely not. Of course, the "intelligence" of computers depends on program software, operating software, clock speed and how processors are organized, not just on the size of the transistors used. We can therefore easily postulate that the performance of chips will continue to double every 18–24 months, up until 2050. This will largely be owed to the occurrence of quantum computing, optical computing or a combination of both.

2.1.2 Past Versus Future

Current ways of humans to connect to the Internet is via an interface (be it a Desktop PC, laptop, tablet, or smartphone). The interface transforms the input of the user (which is usually given via an input device such as a keyboard—hardware

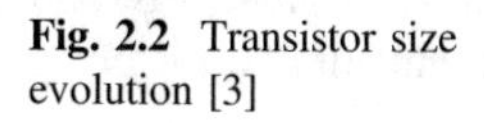

Fig. 2.2 Transistor size evolution [3]

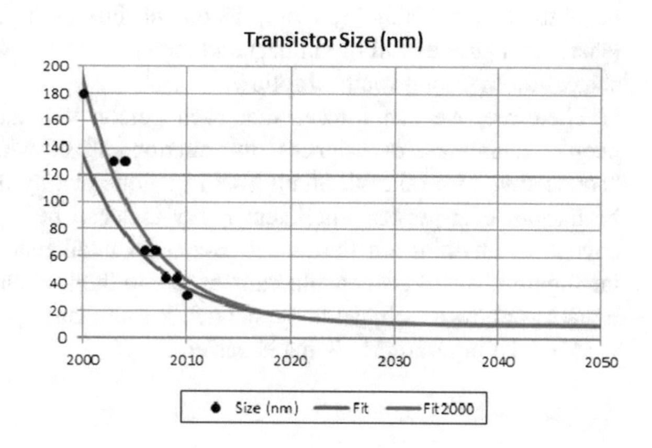

or touch) into information that is then transmitted via a communications medium through the network of interconnected nodes (a.k.a the Internet). At the receiving side, depending on whether the user wants to retrieve some knowledge, or wants to communicate with a second user, another interface will either retrieve the information and send it back to the user, or it will transform the information into output for that second user (which is usually rendered on a screen of different sizes).

Will this way of connecting to the Internet change? Probably so. The flow described above is valid for individuals without impairments. The ones that have visual impairments, physical impairments, or even hearing impairments have to follow different paths usually involving some screen reader or other ways of inputting information. Also inputting information follows a long route, in which the brain first thinks what to input, then it commands the hand to do the motion(s) through which text is written or a button is clicked. We can easily observe that advancements (which are usually derived from the need of overcoming current human limitations) might lead to some other way of inputting and rendering information.

We can foresee that there will be no such things as phones, tablets, or computers. Instead computing will just be embedded into ourselves. There will be no need for keyboards or screens for inputting or rendering information, saving the unnecessary steps taken from the thought to the actual input. Inputting information will be as easy as thinking. The same will happen with displaying information. The visual information will not reach our brain through our eyes, but it might arrive directly at the brain, which is permanently connected to the Internet.

Given the above considerations, we can assume that there will be no such thing as "connecting to the Internet". We will be on the Internet at all times. In fact, every living and non-living thing on the planet will be on the Internet at all times. This will also mean that all of their data will be recorded onto the Internet which might have different implications which will be analysed in Sect. 2.3.

Nowadays we keep collections of pictures onto hard drives or online storage services, we keep people in digital address books on smartphones or social network accounts, we keep information about ourselves on social networks or in personal (digital) documents and, finally, we have memories that we keep in our mind. In order to share a "thing" (picture, document, link etc.) you need to press a button (share, compose e-mail or similar) and then choose via which channel you want to share. Another long route to follow.

Therefore, we can foresee that each person will have its own collection of people, memories, experiences, information, all of which will be shareable by "connecting with others". Sharing will be done just by thinking. The Internet will be like an ever-present intelligent *entity* that can be tapped into at any time by anyone or anything on the planet. Being an intelligent entity driven by artificial intelligence, it will also enable humans just to think of an outcome and the Internet figures out how to achieve that outcome. It's similar to programming today, but the work of the programmer is made easier.

2.1.3 Ubiquitous Connectivity

To enable what has been described above it is most likely that at a fairly early age we might have tiny, nanoscale devices placed in the body. These will be small transmitters for connecting (actually *being connected*) wirelessly to the Internet. But there will also be sensors. These would have the role of interpreting brain signals, but also the role of performing diagnosis of the health status of the human. There will be real time big data analysis sent to medical centres or other similar entities which will enable earliest possible detection of disease to ensure optimal medical tracking.

Items purchased will, by default, have Internet connectivity enabled. Therefore, any prized item will not be able to be lost or stolen, because it will be always connected to the Internet, and will be able to be tracked. We might even get its location instantly in our brain. This is something that will certainly be a reality if we think about what is happening nowadays. It only took 10–15 years' time to dismiss the idea of the lost friend. You can, most likely, find him via a social network or via Google search.

Homes, and, more generally, buildings will be made smart by the cluster of web enabled devices that come and go within the home. The devices inside the buildings will perform different roles and will enable the building, overall, to be *intelligent*. The devices will interact with each other, driven by the *intelligent* Internet, in order to perform different roles and contribute to the intelligence of the entire building. For instance, the building will be able to have self-sufficiency, self-optimization and self-healing. Devices in the building will be able to create their own social networks within a space.

Not only building or other structures will be self-healing. The entire web will develop in resiliency and mobility. The (core) network nodes will no longer be limited to being fixed and the traffic limited to going through a limited number of routes. There will be devices that come and go inside a network in a plug-and-play manner, and the higher layers in the protocol stack will be oblivious to how the physical layer is made up.

All of these scenarios will be enabled as mentioned above by the proliferation of nanotechnology and optic computing, and also by advances in brain science. It will also be complemented by such concepts as software-defined radio, virtualization, software-defined networking etc.

2.2 Industry and Economy in 2050

Again, in order to predict economy and industry, we have to look back at the past and present. If the 50s through the 70s brought about the third industrial revolution [4], which was the change from analog, mechanical and electronic technology to digital technology, current times (2010s) are bringing about the notion of *Industry*

4.0 [5] or the fourth industrial revolution. This is a term that embraces a number of automation, data exchange and manufacturing technologies. It can be viewed as 'a collective term for technologies and concepts of value chain organization' which draws together current concepts such as Cyber-Physical Systems (CPS), the Internet of Things (IoT) and the Internet of Services (IoS). We see the use of Information and Communication Technologies (ICT) in all fields, starting from industry, and going to labour, health and, most of all economy.

We can postulate that the information and industrial economy will have joined up entirely by 2050. Almost all economic activity starts at ICT work which creates intellectual property (IP). The main human activity will possibly be Computer-Aided Design (CAD) and design work. The creation of physical things will be done by all-purpose or special machines, similar to what 3D printing is producing nowadays. Therefore, there will be an *idea-driven economy* as opposed to a *production-driven economy*. Due to the instantaneous nature of communicating and connecting, the economy will be profoundly global on ideas. People from any corner of the planet will take part in the knowledge sharing and idea producing activities. Also, the economy will also be local on production, since a physical device will be able to be produces anywhere.

This is the opposite of what is happening nowadays, where production is done in different facilities across the globe (economy is global on production) and then products are shipped to different destinations, and ideas are produced by central teams or management (economy is local on ideas).

Job schedules will be made up of intense bursts followed by days of downtime (you could think of this in terms similar to today's medical residents). People will not be able nor will want to work 8–9 h continuously, under a strict schedule and then resume the next day. This is becoming more and more prevalent right now, because of the amount of social networks that many browse during work hours, and because company e-mail and computing is becoming available on personal smartphones, and will only continue to rise in the future. Only some critical task can keep employees focused or busy from the temptation which is social networking. Process automation will therefore have to focus on maintaining a near constant productivity despite alternating intensity of work effort. We can imagine there will be such scenarios of job sharing, where a job is made up of different small tasks that an employee can carry out in a short burst of productivity, but also job or task offloading, where a critical task can be carried out in an automated fashion, and the decision to offload is taken on-the-fly.

Working remotely also might become the norm. It is enabled just by the mentioned advancements, such as instantaneous connecting to the Internet and sharing of thoughts, ideas, and memories.

There will be new types of jobs for new type of economy and industry. The key among these skill sets will be the transition away from management towards leadership. The manager will be the leader of a set of people in the Internet. If nowadays, the manager's role is focused more on organization and planning, he will transition to that of a leader, where a leader's role is to inspire and motivate. The critical tasks of organization and planning will be done in an automated way, optimized and taking into account all possible job-related metrics.

2.3 Losing Access to Technology. Security and Privacy Aspects

Today, technology and, especially, access to Internet, is already entrenched in many of the people of this planet. It is not quite uncommon now for people to become restless if they do not have access to their Internet-enabled mobile phones (e.g., nomophobia [6]). Also, there are scientific studies that link Internet addiction to modifications in brain structure [7]. We can envision that, without the Internet, in 2050, one is functionally useless and at a great disadvantage when earning income or fostering relationships.

Imagine, therefore, that in 2050, a criminal sentence might become to lose access to technology. We can imagine that, similar to drug or alcohol addiction, the lack of Internet use might bring side effects such as anxiety and stress.

Criminals, the unintelligent, and delinquents might, therefore, make up the majority of unskilled jobs where physical presence is required. They will be thought of as second class citizens and there will be something that can be called a digital divide between the so-called upper or middle class citizens and lower class.

There could, also, be, people that willingly choose to be "disconnected", because of not wanting to be tracked. There will, likely, also be a proliferation of the dark web where more and more people (either criminals or people that simply do not want to be tracked) will take refuge from tracking that could be done by corporations or governments.

So, to summarize, in 2050, not being connected might be punishment for committing crimes, or it might be a choice done by some people, whether criminals, or people that do not want to obey the rules.

However, losing access to technology might happen not by a sentence or by voluntary choice, it might happen as a result of a technological threat or virus. Worse, connectivity might be lost between machines that perform critical tasks. We know that, nowadays, security is ensured by different measures such as encryption and authentication or maintaining certificate authorities. What will security measures look like in the networks of the year 2050?

We have already determined that a variety of intelligent systems, operating with varying and controlled degrees of autonomy, will continue to proliferate by 2050. Sensing, communicating, collaborating intelligent entities will densely populate the technological space exhibiting a range of sophisticated capabilities. We have to think also that nations, rogue groups, and malicious individuals will step up their hacking games. The hacks could affect banks, businesses, and private data, but also do tangible damage to a world increasingly reliant on technology. Citizens will most likely divide between those who prefer convenience and those who prefer privacy, as is much the case already today.

We can most likely say that security will be embedded in devices from the design stage, and it will be compulsory. Every device that is manufactured will have natural cybersecurity. Also, threat and vulnerability information will be shared and coordinated across devices in order to increase the security and privacy of devices and humans.

Regardless of all these, there will be an exponential increase in the number of cyber-attacks, but there will be fewer such successful attacks than we have now. However, the attacks will be much more devastating, with the ability to bring to a standstill even an entire country. There will be so-called "threat prediction" mechanisms in place, but, as is the case with security nowadays, a threat will always appear and wreak havoc before the solution to this threat emerges.

2.4 Open and Closed Systems

Looking at the previous sections, we have to think about the Internet and systems in general how much openness can be preserved.

We can say that a major part of the innovative value of today's systems stems from the openness to novelty and experimentation. As we use it right now, the Internet means not having to ask anyone for permission or (sometimes) pay to access a resource or information.

There are some companies committed to preserving openness, releasing open source software that can run on different types of hardware. There are also others committed to closing it down, by requiring special hardware for its closed source software to run. Each parties have their motivations for this, ranging from ideology to user experience, profits etc.

We can however, from these observations, outline two distinct evolutions, or two distinct futures of the Internet.

The best scenario is that there will continue to be open systems and their advantages (but also disadvantages) can be highlighted in comparison with closed ones. Their advantages come from the open features and experimentation that can be performed, but also disadvantages like lack of interoperability between different producers, different user experience for different hardware etc.

Another scenario is when all systems become closed, and cannot operate in open way. There might be different reasons for that. It could come through government intervention or as a threat from viruses. It could even come from the public pressure. Technological threat (viruses) as a result of the existence of openness might become unbearable for the average citizen and the system, at the request of the public, can be closed down. Meaning all applications have to go through a censorship committee or no application can be released without satisfying some criteria possibly having to do with it being perfectly compatible with the rest of the system.

Table 2.1 Comparison between open and closed systems

	Open systems	Closed systems
Support	More prompt, less specialized	Less prompt, more specialized
Customization	Available	Possible, with likely breach of terms and conditions
Security	More prone to threats, but security bugs can be fixed more rapidly	Less prone to threats, security bugs are fixed in slow update cycles
User privacy	Less	More
User experience	Different experiences for each user, non-tech users may have bad experience	Users have similar experiences, even the non-tech savvy users
Operator financial gain	Less	More

We can argue that the average citizen has zero ideological commitment to the openness, only to utility. And this could be used as a pretext for restricting the openness of technological systems (Table 2.1 presents a comparison between open and closed systems in terms of several key parameters).

2.5 Discussion: Two Future Worlds

Taking into account the two scenarios outlined in Sect. 2.4, we can say that there will possibly be two worlds. There might be a world where open systems continue to exist. They are ways of empowering groups and societies, marginalized groups. Take, for instance, a mobile application that enables citizens in a community or city to report problems in that area. It is produced by a developer and released in an application store and downloaded and used by hundreds or thousands of people. This puts pressure on the leadership of the city to take actions against those issues. This wouldn't have been possible if the application was subject to some audit and it might have been stopped from being published.

Opposite to this, there might be a world where everything is closed and people know only what large stakeholders (e.g., governments, large corporations) want them to know. People that believe and accept without question, what is being served to them. It can be clearly stated that there is no easy way of dealing with the informational deluge, filtering important topics of interest from all the heap of distractions that are served via social media or other channels. It is what is happening even today, and it will be happening on a larger scale in 2050. Many people on the planet have become content producers, either low quality or high quality. And the advent of social media has enabled the content that is produced to be instantaneously shared with a large mass of people.

2.6 Conclusion

Part fact and part fiction, the present paper has been written as an exercise of envisioning how technology advances and how the world will look like in 2050. It is based on analysis of current trends, advances in technology such as Moore's Law, future medical breakthroughs and much more. Where possible, references have been provided to support the predictions.

The analysis has outlined two different scenarios that might become a reality in 2050. One scenario is that there will be a closed world, where people only know what stakeholders (governments, large corporations) want them to know and are served specifically tailored content, transforming them into informational "couch-potatoes". They have access only to the systems and information sources that they are allowed, and any content that is produced has to be approved. The other, more optimistic, scenario is a world where open systems continue to exist and serve as means for people to become more actively involved in the problems of society, empowering different marginalized groups and societies.

References

1. Moore GE (1965) Cramming more components onto integrated circuits (PDF). Electron Mag 4. Retrieved 11 Nov 2006
2. Projections. Online. http://www.futuretimeline.net/21stcentury/2050-2059.htm (Available)
3. Roper DL (2010) Silicon intelligence evolution. Online. http://www.roperld.com/science/SiliconIntelligenceEvolution.htm (Available)
4. The Digital Revolution (2008) Online. http://web.archive.org/web/20081007132355/http://history.sandiego.edu/gen/recording/digital.html (Available)
5. Schwab K (2016) The fourth industrial revolution. World Economic Forum
6. Bianchi Adriana, Philips JG (2005) Psychological predictors of problem mobile phone use. Cyberpsychol Behav 8(1):39–51. doi:10.1089/cpb.2005.8.39
7. Young KS (1998) Internet addiction: the emergence of a new clinical disorder. Cyberpsychol Behav 1:237–244

Chapter 3
Beyond the Next Generation Access

Vladimir Poulkov

Abstract This chapter considers the telecommunication access network technologies and their development trends in the future. The impact of these technologies on society is pointed out and an overview of the fixed and wireless Next Generation Access (NGA) is presented. The evolution of the current access networks, including topics such as future fiber and wireless access technologies and architectures, convergence of fixed and wireless access and virtualization of the core network are discussed. Based on the review and analysis of these issues, the idea for the development of the future telecommunication networks towards a Unified Virtual Network, having a unique and unified core and access infrastructure, is presented. A vision for the future development and evolution of such a virtual network is outlined, and the way that the physical and theoretical limits of the communication link throughput could be reached and even in some way exceeded. The development and evolution is considered to be a constant process towards super intelligence and perfection in such a way that the access network with all its users (humans and machines) will become a very intelligent and unique entity.

The rapid advances and evolution of telecommunication technologies are moving the world toward a fully connected or *Networked Society*, where access to communication resources and exchange of information, will be possible anywhere, anytime, by anyone or anything. "*From the smallest personal items to the largest continents, everything everywhere will be digitally connected and responsive to our wants and likes*" [1]. Today's vision for the access network of the future implies that the future access technologies, besides people, will support connectivity of machines and devices, and a diverse range of all other things that may be connected. To manage all these connected units, and the increase in traffic demands, the future generations telecommunication access systems will need to implement novel functionalities and support a much wider range of services, thus bringing up new requirements to the way

V. Poulkov (✉)
Technical University of Sofia, Sofia, Bulgaria
e-mail: vkp@tu-sofia.bg

R. Prasad and S. Dixit (eds.), *Wireless World in 2050 and Beyond: A Window into the Future!*, Springer Series in Wireless Technology,
DOI 10.1007/978-3-319-42141-4_3

telecommunication networks are accessed. If access networks are to handle this constantly increasing traffic load in an affordable and sustainable way, throughput and Quality of Service (QoS) must increase dramatically, while cost and energy consumption need to be radically lower than they are today. Practically, what is expected to happen is a radical change in the infrastructure of the access network, driven by technological development and by the way people and machines will communicate. A need for change in the current telecom architectures and infrastructures could be recognized today, as even nowadays driving forces that are strong enough to push forward to such a change could be defined. The provisioning of higher data rates will continue to be a key driver in network development and evolution, but there are also other important drivers related to the manufacturers of new devices, the developers of new applications and services, customer demands, business competition, costs, complexity and efficiency. The manufacturers of new devices and the developers of new applications constantly release newer and smarter user devices, new applications and application-driven platforms and services. These novel communicating devices and machines, when introduced into the network, bring up more and more demands and new requirements on the current infrastructure. The Internet enables the everyday generation of new applications and services, without any fundamental constraints to deliver them anywhere in the world to any user using any device, thus requiring from the telecommunication networks higher and higher transmission speeds.

The development of Internet of Things (IoT) and Machine-to-Machine (MtM) communications could be hardly supported by todays' telecommunication infrastructures. In addition the type of traffic in the network is changing. There is an enormous increase of the multimedia traffic as applications, social networking, gamification, etc. are surpassing pure voice services and which together with all the novel types of communications imposed by the IoT, will raise more and more problems related to the ability of processing the volume of data transmitted by end users and devices especially in larger urban areas.

The overall pricing of services and cost models of the networks infrastructure are also under consideration. Even today the majority of the users are not concerned about technological details. Their main concern is to obtain cheaper and cheaper services that are not constrained by time and location (i.e., anytime and anywhere). On the contrary, the requirements towards secure, reliable and efficient data transmission, data management and QoS are growing, as this from the operators point of view, is a prerequisite for market domination. This of course requires good and efficient Operation, Administration, Maintenance and Provisioning (OAM&P), which brings up the necessity of more reliable and easier for maintenance access infrastructures. The support for mission-critical MtM communications and applications will require ultra-reliable connectivity with guaranteed availability and reliability.

Pressure for changes in the overall model of the infrastructure and architecture of the different types of networks are also lower costs, efficiency (higher throughput, lower power consumption) and green issues such as electromagnetic compatibility and health. Built on a vision of massive machine-type connectivity, with tens of billions of low-cost connected devices and sensors deployed, the future *Networked Society* will require the mass-availability of truly low-cost devices. These devices will not only need to be affordable, they will also need to be able to operate on battery for several years without needing to be re-charged, implying to meet the strict requirements for low energy consumption.

In short, at any given time now and in the future, next-generation access networks will have to be able to ensure ultra-high data rates, the connection of massive numbers of devices, very high volumes of data transfer, low cost devices, ultra-low energy consumption, and exceptionally high reliability and QoS. The chapter is organized as follows. In the following point is explained what is Next Generation Access (NGA) and its societal impact. The current access technologies are reviewed and their limitations outlined in point 3. In point 4, the development and trends of NGA are considered. The final section of the chapter is related to the expectations of the access networks beyond NGA.

3.1 Impact of NGA on Society

Short time forecasts state that the number of devices connected to IP networks will be three times as high as the global population in 2019 [2]. There will be three networked devices per capita by 2019, up from nearly two networked devices per capita in 2014. Accelerated in part by the increase in devices and the capabilities of those devices, IP traffic per capita will reach 22 GB per capita by 2019, up from 8 GB per capita in 2014. With such a forecast, the NGA is considered today in the context of delivering broadband services at a speed of 30–100 (200) Mbit/s to subscribers. In most cases this is related to the integration of fiber optic networks with the next 5G mobile communication networks or Fiber-Wireless (FiWi) networks. FiWi networks are based on the convergence of various optical and wireless technologies, exploiting the advantages of their complementary features and realizing a communication infrastructure combining the reliability, robustness, and high capacity of optical fiber networks and the flexibility, ubiquity, and cost savings of wireless networks. The combination of both wireless and shared passive fiber media is expected to support a high speed access of even more than 100 Mbit/s and will help the realization of adaptable, dependable, and energy-efficient broadband access networks with advanced management, reliability and survivability techniques [3–6]. Ensuring such a high-speed subscriber access is considered to have a crucial impact on economy, industry and the whole society.

The impact of ensuring NGA, as considered nowadays, is strongly related to the economic and social development of the society [7]. Access to adequate broadband services has crucial importance to the economic and social development. NGA has an overall impact on:

- *Jobs*. Enterprises with strong web presence are drivers of innovations and job creation.
- *Competitiveness*. Advanced NGA services are crucial for productivity growth, improvement of competitiveness, the creation of new opportunities for entrepreneurship and jobs.
- *Attraction of Investments*. The availability of NGA will enable the most advanced uses of ICT and investments from global corporations.
- *Healthcare Reform*. e-Health technologies such as remote monitoring and remote diagnosis provide a tangible opportunity to shift the balance of health-care away from the hospitals and into the community and even home.
- *Transport*. Use of NGA for increased e-Working and interactive traffic management will reduce peak traffic flows, impacting positively on energy use, carbon emissions and efficiency.
- *Education and e-Learning*. NGA will provide a platform to transform the educational experience by bringing dynamic resources into the classroom and enabling seamless communication between teachers, students and parents.
- *Citizens, Consumers and Government*. NGA will give all citizens access to the same information and opportunity regardless of age, class or location.
- *Regional Development*. Deployment of NGA will help to resolve many of the key issues associated with remote regions thereby enhancing the local productivity.

Society has changed in many ways over the last century and will continue to change at similar rates over the next centuries. Information and communication technologies and electronic devices are drastically changing our lives, society, and economy. Since some of the changes in the last century have been enabled and driven by specific technologies, some future changes would also have this character [8]. Much of our interaction with the world is mediated today by communication devices such as smartphones and other smart devices and machines. They are increasingly becoming our major source of knowledge and interaction with people and environment, as well as our major means of social life. In the next decades, population, economic and social trends will very likely be shaped by this process and will tend towards a hyper-connected world of machines and humans and to a transition towards individual societies and economies based on novel information and communication technologies and machine intelligence. Hyper-connected telecommunication infrastructures and networks will be the "*single, most indispensable element of binding humans and individual societies, industries, economies, and humans, building an infrastructure of large-scale, complex and highly networked systems whose efficiency, sustainability and protection would require intelligent, interoperable and secure ICT solutions and novel business models*" [9]. Hyper-connectivity will be one major symptom of progress resulting in our world becoming progressively more connected at many levels: starting from global transportation networks that carry people and goods across the planet and culminating into global communication networks that mobilize and distribute information at an ever increasing volume and speed. A hyper-connected world is a world where every agent is connected by numerous means to many other agents and where distances, both in space and time, are collapsing.

A hyper-connected world will have a different societal impact in addition to the issues mentioned above. Hyper-connectivity will have huge implications for society, for businesses, for consumers and for the very structure of the global economy. Besides, that it will change the way people interact with each other, such a hyper-connected society will change fundamentally education, medical care and well-being, transportation, energetics and energy-efficiency, infrastructure, human activity, public services, etc. It will have an enormous impact on economic growth, and will reshape a lot of industries, introducing full automation and smart manufacturing concepts, and will disrupt many others [10, 11].

Having this in mind it is worth to mention that the present day vision of NGA will not fit in such a future hyper-connected scenario with its economic and societal impact. The focus of NGA is more or less in the way as planned today, it is still on technical connectivity, routing, networking, OAM, control, etc., based on the principle of building the network as a connection of uncoordinated communication infrastructures with distributed and relatively little intelligence. The tendencies towards Self Organizing Networks (SON) is one step forward to a novel approach towards incorporating intelligence in the networks, which will have influence also on the way the communication networks are accessed. SON is a concept introduced by the 3rd Generation Partnership Project (3GPP) as a step towards achieving Cognitive Networks and Autonomic Networking in the future, thus targeting the creation of self-managing networks which will enable further growth with less or none human intervention [12–14]. For sure NGA will also implement some of the intelligence and the cognitive and autonomic networking features foreseen for SON; as for the realization of a hyper-connectivity access infrastructure the intelligent provision of resources, i.e., super-high speeds or super-broadband access will be a must [15]. Starting with the current access technologies the development towards such an access infrastructure can go through different scenarios, the major of which are considered below.

3.2 NGA Technologies

Next Generation Access networks are considered as an essential element in ensuring fast broadband (>30 Mbit/s) and ultra-fast broadband (>100 Mbit/s) access, for the provision of services with increased performance, improved quality of service and symmetry of speeds in the up and downlink. In practice, NGA networks are characterized by enabling significantly higher speeds of access than the "basic" broadband (>2 Mbit/s) provided by many of the networks today. Current access technologies and the potential speeds that they can achieve are given in Table 3.1. It could be seen that for the implementation of NGA the fixed optical access based technologies of Fiber to the "x" (FTTx) are the ones that mostly conform with the NGA speed requirements.

Table 3.1 Potential speeds of current access technologies

Technology	Throughput
Fiber to the cabinet (FTTC)	40–100 Mbps
Fiber to the premises (FTTP)	70–100 Mbps
VDSL	25–100 Mbps
Data over cable service interface specification (DOCSIS) 3.0 technology	Higher than 100 Mbps
Fixed wireless	1–100 Mbps
Wireless mobile/mobile broadband	up to 30 Mbps
WiMAX, long term evolution (LTE)	up to 100 Mbps
Satellite	up to 10 Mbps
5G (pilots)	higher than 1 Gbps

3.2.1 Fixed Access Technologies

In Fig. 3.1 are illustrated the different architectures for the realization of FTTx networks. In the point-to-point (PtP) architecture all subscribers are connected to the Access Node (AN) by individual fibers. For the realization of this architecture are necessary a large number of fibers which increases the installation costs and maintenance costs. In addition, each connection requires two interfaces, which leads to an increase of the hardware and overall energy consumption. For the reduction of the number of fibers in the access network point-to-multipoint architectures are used. Such architectures introduce one or more additional aggregation layers between the subscriber and the local node (station). As seen from Fig. 3.1 in an Active Optical Network (AON) the network functions (second and third level of the OSI model) are delivered closer to the customer. Most often the AON networks are characterized by an active aggregation element (e.g., Ethernet switch or hub) in the last mile of the network. Different solutions for such a hub to be located in a street cabinet or in a building next to the subscribers can be proposed. The AON allows a reduction in the number of fibers as compared to the PtP, but the number of interfaces (i.e., hardware and network power consumption) is not reduced.

Unlike AON in the Passive Optical Network (PON) the aggregation is performed based on passive optical devices and components, such as optical splitters or Wavelength Division Multiplexing (WDM), which means that the passive architecture enables an optimization and reduction in the number of fibers, hardware and energy consumption. In PON the optical cable is installed between the optical fiber line terminal equipment and a remote terminal unit (usually an optical power splitter) located in the service area (up to 20 km from the central office). From the remote node, the subscribers or the optical network devices are connected via optical splitters [16]. The existing PONs (Ethernet, Gigabit) typically use two different wavelengths as channels for duplex transmission over a single fiber. The downlink channel (1490 nm wavelength) is inherently a broadcast channel and in this case each optical network device filters the received data. The uplink channel

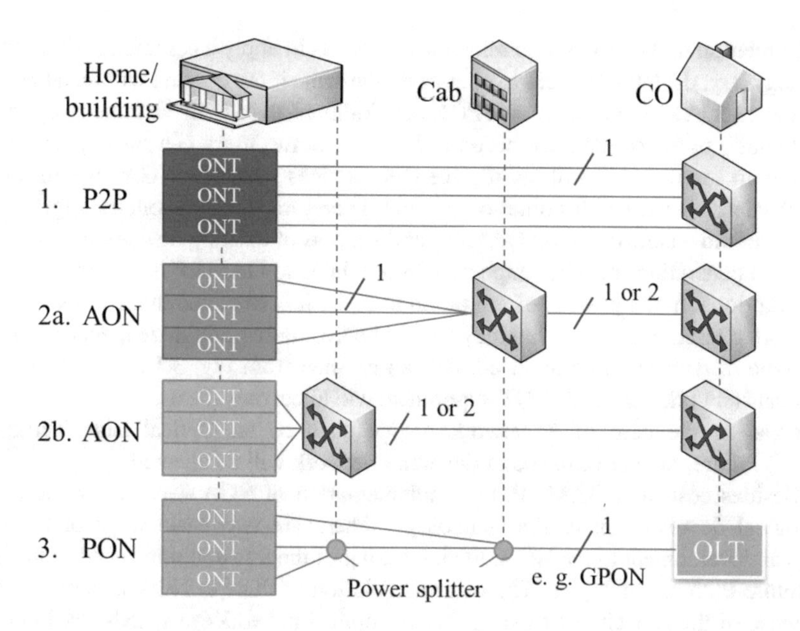

Fig. 3.1 Architectures for the realization of FTTx networks

(1310 nm) is shared for all optical network devices. Time Division Multiplexing (TDM) together with a dynamic algorithm for frequency band allocation is used for the provision of various services to users. Therefore, these networks are called TDM PONs [17–20].

3.2.2 Evolution of Passive Optical Networks

NGA PON may evolve in different ways depending on the requirements imposed on them. There are basic requirements, which determine their development [21]. These requirements are economic (related to investments and/or profits) and operational (OAM&P and/or QoS). For the migration towards PON new investments will be required, as the necessary application and use of a new technology in addition to the existing, or replacement of the latter at the end nodes (last mile) of the network, will be the obvious approach. Thus the minimization of investments related to the equipment is an issue which will be taken into major consideration in addition with maximizing the profits from existing resources and reuse of the existing optical infrastructures. An example is the efficient use of the network capacity through the application of dynamic resource management approaches (allocation of frequency bands or wavelengths), which results in better revenues and a faster return on investment.

In relation to the OAM requirements, the idea is to apply a common maintenance strategy, i.e., the NGA devices to operate on the same infrastructure without affecting the current OAM. In the process of PON development, even with the same category of customers, traffic needs may become different. Some users can be satisfied with minimum services and will not replace their devices with new NGA, or will do so much later, when prices become comparable. Therefore, development (or upgrade) to NGA should enable common OAM&P and support of existing devices and those of the next generation. Another important issue related to OAM&P is the avoidance of interruptions. In the process of migration to NGA it is expected interruptions in the network operation to occur, but they should be avoided or minimized, depending on the type of devices to be replaced. As can be seen from Fig. 3.1 an outage of the Optical Network Terminal (ONT) equipment will affect only the users connected to it, whereas in the case of a network device such as an Optical Line Terminal (OLT) failure, the performance of the entire network will be affected.

Besides costs and OAM&P, the implementation of NGA depends on the technological development of the technology. There are two basic ways of technological development to the NGA: the increasing of the transmission speed or the use of future PON technologies. The natural evolution of PON to NGA is based on the increase of the capacity of existing passive optical networks to speeds reaching up to 10Gbit/s. In practice, there are already standards for 10Gbit/s next generation PON. The currently developed standards for this type of networks are influenced by the possibility of parallel operation with the existing passive optical networks, the prices for installation and maintenance and the ease of implementation.

In September 2009 IEEE ratified a new standard for 10Gbit/s EPON (IEEE-802.3av). ITU-T (Question 2, Study Group 15) issued a series of recommendations for 10Gbit/s-GPON (XG-PON), basically G-987.1, G-987.2 (both of them approved in January 2010) and G-987.3 (approved in October 2010). Both recommendations, (IEEE-802.3av and ITU-T), which offer NGA architectures, are a good example of the increase of the transmission speed allowing compatibility with the existing old-generation PON networks [22]. In the longer term PON is expected to evolve and reach speeds of the order of 100 Gbit/s. However, for higher speeds, it is difficult to reach the typical distances for PON networks without amplification of the signals. This evolution or migration can be done on the principle "in case of need" and two phases of evolution are foreseen: asymmetric and symmetric speed upgrade [23].

In asymmetric speed upgrade, the downlink traffic is usually higher than the uplink. In this case PON is attractive because of its ability to broadcast the signals on the channel downwards in the direction of the user. Another reason for the asymmetric migration is the fact that the addition of a transmission opportunity to a 10Gbit/s uplink (symmetrical approach) results in the need for installing more expensive fiber optic network devices. One example is the introduction of devices for WDM [24, 25].

In the case of symmetrical upgrade, the speeds in the two directions of transmission are aligned (for example to 10 Gbit/s), depending on traffic needs (e.g., multimedia services) and the number of users connected at the terminal points.

Practically there are two ways to upgrade the speed symmetrically: by TDM and WDM [26]. In the first case, the increase of the transmission speed in the uplink is achieved by applying a method of time sharing of one wavelength and the use of two different transmission speeds. This approach was approved by the IEEE for 10Gbit/s EPON. It leads to a reduction of the installation costs, since the existing uplink channel operates in the lower wavelength. Newer optical network devices work with distributed feedback lasers and in this case they that can be integrated into the existing infrastructure, thus reducing costs. But here the practical implementation becomes more complex because it is necessary to introduce an additional mechanism in the network for the management and control of the various transmission speeds and for synchronization [27, 28].

In the case of symmetrical upgrade based on WDM are added channels at a speed of 10 Gbit/s operating at different wavelengths. This approach may be more expensive than the asymmetric, because the transmission medium and systems cannot be fully reused. An example of this is the need to work in other wavelength bands, such as C or L, but they can already be reserved for use for analog or digital video broadcasting [29, 30].

For the migration to NGA some new or future technologies that are not yet fully standardized or explored could be applied. These technologies are based on the use of novel multiplexing methods, such as CDM (Code-Division Multiplexing) and SCM (Sub-Carrier Multiplexing), or coherent PONs [31]. By using different wavelengths for different PON generations, a hybrid structure could be achieved in which different PON generations operate independently of each other. Such hybrid networks may be based on CDM or the so-called OCDM-PON (Optical-CDM PON) in which the increase of the capacity of the system is a result of the introduction of Code Division Multiple Access (CDMA) [32]. Other examples of the implementation of NGA PON networks are based on Orthogonal Frequency-Division Multiplexing (OFDM), the so-called OFDM PONs or coherent PONs, which use coherent lasers for operation in ultra-dense-WDM (U-DWDM) mode [33–35]. The development of the PON technologies in the future with some of their features is illustrated in Fig. 3.2.

3.2.3 Wireless Access Technologies

As mentioned above, NGA is expected to be based on a combination of wireline and wireless technologies. The access network topology will be a mixture of wireline and wireless access systems, point-to-point and point-to-multipoint connections, with the inherent pros and cons as shown in Fig. 3.3 [36, 37]. But current wireless technologies such as WiFi and mobile broadband are not the best candidates to deliver the necessary speeds for NGA. The perspective of wireless technologies for the realization of NGA is dependent on the technological development and especially on the evolvement of LTE-Advanced and the future development of 5G wireless technologies, as illustrated in Fig. 3.4.

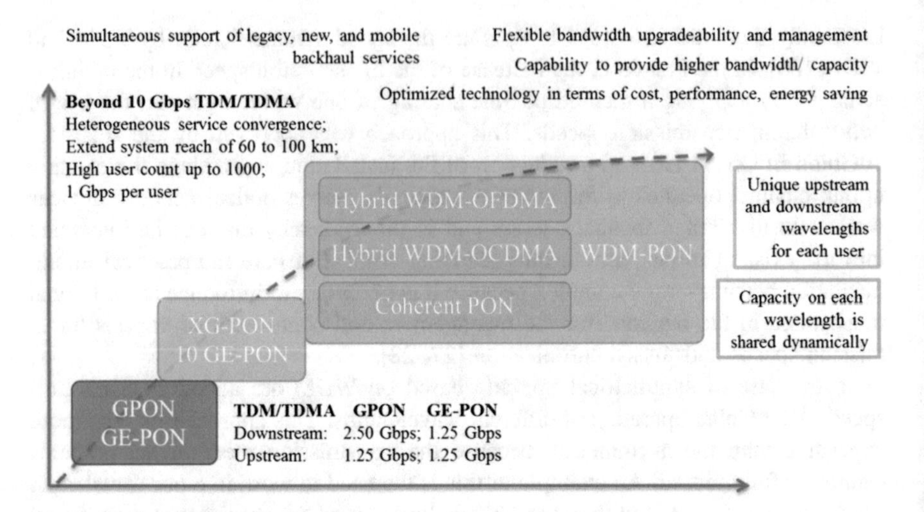

Fig. 3.2 Development of future PON technologies

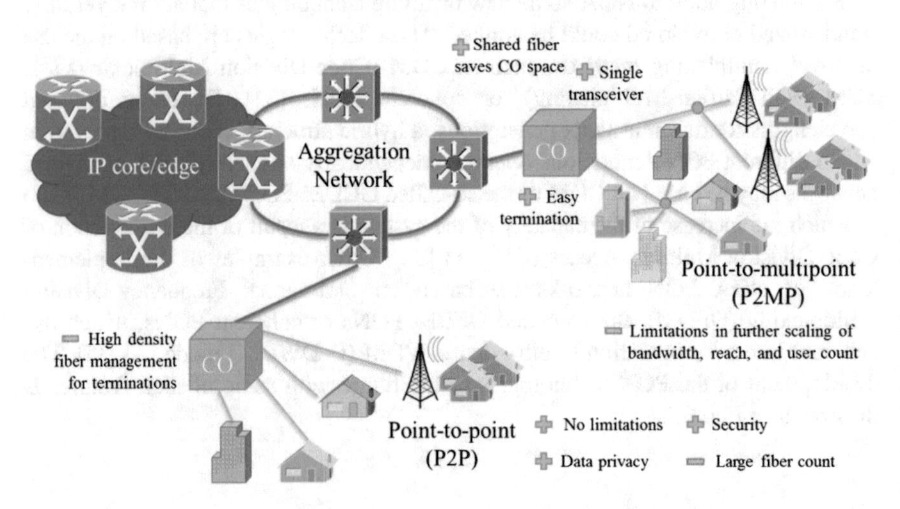

Fig. 3.3 NGA network topology

Regarding wireless technologies, a very important consideration is that the wireless channel is "shared" and practically the connection speed of a subscriber depends on the number of users connected and on the variations of the characteristics of the channel. Therefore, in order to ensure a minimum reliable transfer rate for a subscriber, it may be necessary the wireless access of the NGN to be deployed with some degree of density, advanced wireless configurations and/or novel adaptive and intelligent technologies. Wireless NGA based on adaptive mobile broadband technologies must also be able to provide the required QoS to users in a given location, while serving many other mobile subscribers in the servicing area.

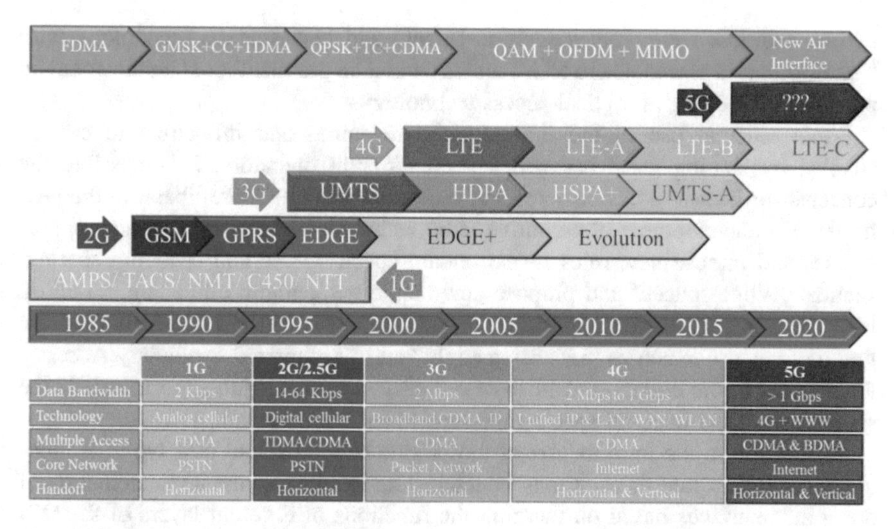

	1G	2G/2.5G	3G	4G	5G
Data Bandwidth	2 Kbps	14-64 Kbps	2 Mbps	2 Mbps to 1 Gbps	> 1 Gbps
Technology	Analog cellular	Digital cellular	Broadband CDMA, IP	Unified IP & LAN/ WAN/ WLAN	4G + WWW
Multiple Access	FDMA	TDMA/CDMA	CDMA	CDMA	CDMA & BDMA
Core Network	PSTN	PSTN	Packet Network	Internet	Internet
Handoff	Horizontal	Horizontal	Horizontal	Horizontal & Vertical	Horizontal & Vertical

Fig. 3.4 Evolution of wireless technologies

If we look ahead and assume that until 2020 the technology "LTE-Advanced" will be widespread and technologies such as MIMO (Multiple Input—Multiple Output) and adaptive beamforming will be developed to a level where the investments for their introduction are significantly reduced, then in certain places the introduction of NGA based on wireless instead of optical access can be justified. In fact, at a certain degree of the development of mobile communications, it may no longer be possible to increase the spectral efficiency due to reaching the theoretically possible values and limitations of the operating frequency bands (throughput saturation). In these cases, the only option is to increase the bandwidth of the operating frequency bands and/or moving to higher operating frequencies and densification of the mobile Base Stations (BSs). This of course will bring a big pressure on OAM&P costs of the Access Points (APs) due to the inherent higher installation costs and lower energy efficiency than in PON. Thus, it is expected that the main innovations in this field in the future will be aimed at reducing the costs per bit of transmitted information and the costs of energy per bit. This can be done by the means of change of the cellular infrastructure, innovative approaches of spectrum sharing and of the air interface.

In order to significantly reduce the volume of the signaling information which is exchanged in the air interface between the BS, an attractive approach may become the amendment of the concept of existing cellular networks and the acceptation that access will be performed through heterogeneous terminals served by many intelligent and self-organizing APs. Different functionalities, such as control of the signaling, two-way data transfer and the mobility functions (such as location and tracking), which are now provided by the serving cell can be transferred for the provision of a group of different "cells" or more generally a group of cooperating antennas. Such a concept of development will contribute to a massive introduction of new wireless technologies such as MIMO and new antenna technology with beamforming in places where NGA access cannot be provided, or is inefficient to be provided employing an optical access technology.

More efficient use of the limited radio resources and infrastructure can be enabled by sharing them between several different operators. It is possible the concepts for licensing of frequencies to change or become more liberal in the near future. The development of cognitive radio technologies have opened new opportunities and impose new rules for secondary spectrum use and spectrum sharing, such as “White spaces” and propose new methods for traffic offloading, based on novel resource allocation approaches. Moreover, new patterns for regulation are finding their way, such as “Licensing Light” and “Authorized Secondary Access”, which refer to the collaborative use of a “noisy spectrum” (junk spectrum), or the shared use of the licensed spectrum in agreement with relevant licenses.

In addition to the introduction of new frame structures and new types of modulation in the air interfaces, it may be interesting to consider the implementation of novel air interfaces based on merging the functions of different layers of the OSI model. This can be extremely valuable if it is considered that NGA may be used in the context of MtM Communications. In these cases, increasing the channel throughput could be done by using a bigger bandwidth and the joint application of MIMO with new modulation methods. Research shows that 10 Gbit/s can be achieved with a bandwidth of 200 MHz through the application of MIMO with eight parallel branches and 256-QAM on each branch. Alternatively, with 350 MHz available bandwidth such a speed can be achieved with six branch MIMO and 64-QAM [38].

3.2.4 Limitations of the Access of Today

The fundamental limitations of today’s telecommunication networks, including the access networks, are mainly related to: the need to support multiple technologies, the lack of flexibility and scalability, heavy constraints on network upgradeability (cost, downtimes), business models not flexible enough to allow for large investments.

Starting with the support of multiple technologies, a good example is the existence of many wireless standards such as UMTS, LTE, WiMax, WiFi, Bluetooth, RFID, ZigBee, Packed Radio solutions, etc. In many cases these standards are related to the same or overlapping services and applications. Practically this could be considered as parallel service development, or parallel investment in one and the same application services and functionalities. In addition, the existence of different wireless standards leads to the necessity to install in one device hardware for several different wireless interfaces. The availability of parallel infrastructures and dense deployments of access points in one and the same locations, delivering one and the same services is also obvious today. In some locations infrastructure density is even higher than the device density. This “deployment saturation” results in inefficient and inflexible resource utilization and redundant network infrastructure investments. Besides cost effectiveness, the deployment saturation raises many other problems such as throughput, interference and electro-magnetic compatibility problems. Even with the application of new highly effective modulations, channel coding and interference suppression techniques, in many places the existing architectures have already reached the throughput limit (Shannon limit). The introduction of the concept of Heterogeneous Networks (HetNets) involving coordinated macro and pico- or nano-cell coverage is considered as a means to improve capacity and throughput. Here the problem is that as the number of deployed cells increases, so does the number of cell edges. At these cell-edges end user experience can be significantly impacted by frequent handover, increased failure rate, and lower throughput.

Another conclusion that could be drawn is related to the strong dependence of users on service providers. Practically a service is available only if the user pays "tax" for the infrastructure, which means that if an access to two services is required, which are offered via two different independent infrastructures, the user has to pay two "taxes infrastructure". These are just a few of the limitations of the access networks of today, but besides the technology enablers these limitations will be also driving forces for the future development of the access technologies. The introduction of intelligent approaches to resource management and the implementation of novel access network architectures, based on the principles of virtualization and commoditization of resources, could be the right approach towards overcoming these limitations.

3.3 Future Development Scenarios for NGA

The expectations of NGA up to a horizon of the next 10 years are related to the elimination of some of the current limitations of the access network infrastructures. The trends are towards making them capable of supporting a much wider array of requirements than today and to introduce the capability of flexible adaptation to different "vertical" application requirements. The next generation (5G) access is considered to be the basic "last meter" access infrastructure which will ensure a dramatic growth in user capacity, quality of service, responsiveness, energy efficiency and number of connected devices while keeping a sustainable cost.

The vision is that in 10 years from now, telecom and IT will be integrated in a common very high capacity and flexible 5G based ubiquitous infrastructure, with seamless integration of heterogeneous wired and wireless capabilities, covering a wide range of services for increasingly capable terminals for human communications, and for an extremely diverse set of connected machines and things. The technical directions and enablers of the 5G network are the incorporation of more spectrum (enabling the usage of frequency bands above 6 GHz for ultra-high speed access), the densification of the infrastructure, the application of new air interfaces with spectrally efficient user multiplexing radio technologies and the support of different deployment scenarios [39, 40].

This ubiquitous infrastructure will efficiently support different distributed and centralized deployment topologies with reduced management complexity based on solutions that unify connection, security, mobility, multicast/broadcast and routing/forwarding management, capable of instantiating any type of virtual network architecture. It will also cover technologies like WiFi to use efficiently spectrum management and offload capabilities. The novel air interface technologies will ensure the efficient support of a heterogeneous set of low to high rate services, local and wide area systems, heterogeneous multi-layer deployments, through advanced Multi Antenna Transceiver Techniques, including 3D and massive MIMO beam-forming. The infrastructure will be based on the cooperative operation of heterogeneous access networks integrating virtual radio functions into service delivery networks including broadcast/multicast technologies (terrestrial and satellite based), supporting Software Defined Networking (SDN) and virtualization techniques of Radio Access Network (RAN) functions, providing the environment for multi-BS attachment. It will support the interconnection of numerous devices with different capabilities, with unified connectivity management capabilities, in terms of security, mobility and routing. Coordination and optimization of user access to heterogeneous radio accesses including ultra-dense networks supported by intelligent radio resource management framework will be one of the major characteristics of such an infrastructure. This covers the joint management of the resources in the wireless access and the backhaul/fronthaul (as well as their integration with optical or legacy copper networks), multi-tenancy for RAN sharing (covering ultra-dense network deployments), realization of the "plug and play vision for computing, storage and network resources through the appropriate abstraction, interfaces, and layering" [41].

The way to this ubiquitous infrastructure is *the fixed-mobile convergence* and the introduction of *hybrid architectures for wireless access*, where the general target is the deployment of a global all-IP wireless/mobile network. Fixed-mobile convergence is practically related to the unification of wireless and wireline voice, video and broadband data services through a seamless integration of wireless and fixed networks. Such scenarios are already being implemented, as they bring advantages not only for end users, but also for service providers and operators. From a user point of view, such a convergence will bring unified billing, ubiquitous and seamless connectivity, and access to a consolidated set of services. From an operator point of view there are technical and economic benefits. They are coming from the possibility of a more effective utilization of the resources of the last mile access network and efficient use of the available public and personal, wireless local and metropolitan access infrastructures. These hybrid architectures, called FiWi are the cornerstone of future broadband installations and will be based on the integration of the Next Generation Passive Optical Networks (Ng-PON) with LTE-A HetNets, 5G networks or Cloud Radio Access Networks (C-RAN) [42, 43].

C-RAN is proposed by the China Mobile Research Institute and stands for Centralized, Collaborative, Cloud and Clean RAN and is expected to be a fundament of the 5G mobile networks. This is a type of radio access network architecture in which the baseband processing resources are centralized in order to realize

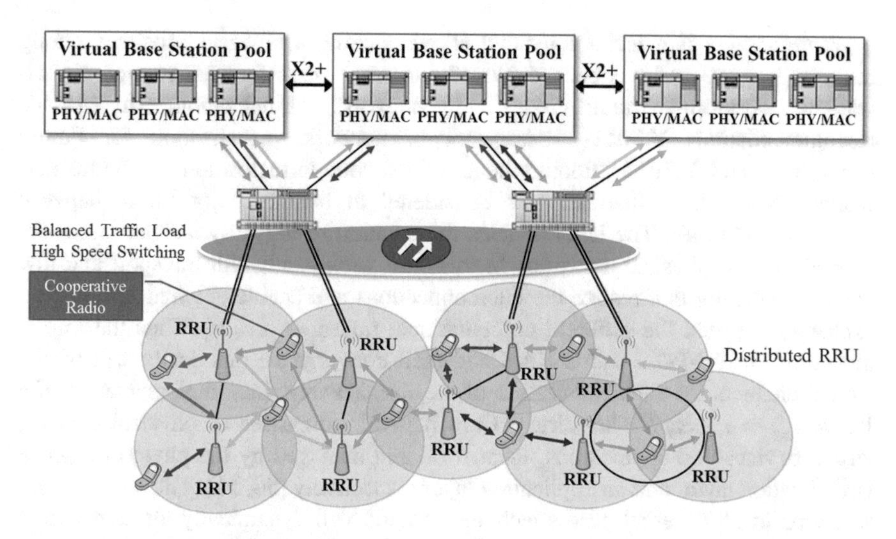

Fig. 3.5 C-RAN architecture

on-demand dynamic resource allocation scheme [44–47]. This leads to an increase in the efficiency of resource utilization and improved spectrum utilization, decreased energy consumption, implementation of interference management approaches in comparison with the traditional mobile BS architectures.

The concept of C-RAN is based on distributed BSs and a cloud based system centralizing the baseband processing resources together to form a pool as shown in Fig. 3.5 [45]. With this approach the resources could be managed and dynamically allocated on demand on the pool level. The C-RAN architecture consists of three parts—a network of Remote Radio Units (RRUs), a centralized pool of Base-band Units (BBUs) and a very high speed transport (basically fiber based) and switching plane providing the connection between the BBUs and the RRUs. The RRUs provide basic wireless signal coverage, while the BBUs are responsible for the dynamic allocation and reconfiguration of resources based on real-time user requests and traffic conditions. Using the concepts of virtualization a Virtual Base Station Pool (VBSP) is realized using cloud computing and software defined radio, where virtualized and centralized baseband and protocol processing, such as PHY and Media Access Control (MAC) layers is implemented. This is one of the major differences between C-RAN and traditional RAN systems where the computational resources are limited within one BBU and are not shared with others. The aggregation and flexible allocation of resources on demand is a feature similar to the cloud and virtualization concepts in data centers. This feature is called resource "cloudification" or "virtualization" and is the core feature of C-RAN leading to improved resource and power efficiency. Through the "cloudification" the BBUs are practically "soft" or virtual, meaning that the BBUs could be dynamically reconfigured and adjusted to increase the resource utilization efficiency.

Virtualization is a technology that allows sharing a physical substrate among multiple systems and has been deployed for many years for data storage virtualization, desktop virtualization and network virtualization. In ITU-T Recommendations Y.3011 "Framework of network virtualization for Future Networks," and Y.3012 "Requirements of network virtualization for future networks", Network Virtualization is considered to be the major future network technology [48, 49]. The latter enables the creation of logically isolated networks over abstracted physical networks. In this case, high bandwidth transport and low latency switching that realize the interconnections and enable efficient information exchange between the different computational nodes, is a must. Thus the implementation of NG-PON and C-RAN and their convergence practically will be the natural enablers for the virtualization of the core network, i.e., implementation of a *Virtual Core concept*. The Virtual Core network comprises of physical devices, virtual devices and applications, located on and managed by the physical layer, a virtualization layer, and an application layer respectively [50, 51]. The Virtual Core will support all different access technologies and will dynamically allocate virtual resources according to the current traffic conditions through its virtual interfaces.

3.4 Beyond the Next Generation Access

To meet future challenges, the expectations from future access architectures are generally related to the development of a big range of the access points (from several meters up to 50–100 km.), ensuring extreme data rates and high traffic capacity for a hyper-connected world of man and MtM communications. This presupposes full integration and virtualization of the fixed and mobile infrastructures and full utilization of the benefits of all the types of wireless networks, which will bring to overall coverage, security, improved services and reduced costs.

From the user perspective, such a full integration and virtualization of infrastructures means that the user will be able to connect to the network from anywhere and at anytime (not depending on the underlying access network) and will effectively utilize the infrastructure (due to the implementation of a shared, flexible and reliable solution). The user will benefit also from the integration of billing and services, as billing (integrated or through bundles), will become only service dependent and will not have a network component associated with it. The introduction and integration of new services must not depend on types of networks or network architectures. Once a service is deployed over the network, it will become a service that can be run from anywhere and at anytime over any type of network.

From technological and service provider perspective, the expectations from future access architectures will give the possibility of modular and flexible territorial spread of the access based on self-organization, self-configuration and self-regulation. This should be a *unified type of dynamic fixed and wireless access* to ensure efficient and scalable utilization of all communication resources. The future access architecture will be cognitive and intelligent, adapting to user and QoS requirements, to new personalized multimedia services and applications, to M2M communications, and at the same time "green" and cost effective, i.e., with minimized energy consumption and very-low-cost deployment and maintenance.

3.4.1 The Virtual Access

The development of access technologies and the virtual core concept and the way towards a future hyper-connected society, lead to the expectation that future access architectures should be considered in the context of the network virtualization. The virtualization of access networks will enable the creation of Logically Isolated Network Partitions (LINP) over shared physical network infrastructures, in such a way so that multiple heterogeneous virtual networks can simultaneously coexist over shared infrastructures (ITU-T Q.21) [52]. The key properties of LINP are: partitioning (each resource can be used concurrently by multiple LINP instances), isolation (the clear isolation of any LINP from all others), abstraction (in which a given virtual resource does need not to directly correspond to its component resources) and aggregation (aggregate multiple instances to appear as a single resource to obtain increased capabilities). Inherent characteristics of such a virtual access network will be the overall intelligence including but not limited to content-oriented networking, energy efficiency, operation based on traffic dynamics, in-system network management, Distributed Mobile Networking and optimized overall performance (Device, System, Network, Path, Network topology).

Major features and functionalities of the virtual access will be related to abstraction, topology and user requirements awareness, self-organization and self-configuration, resource and user discovery, mobility, channel virtualization.

Network abstraction, or the definition of an abstraction level, will allow a selective exposure or hiding of the underlying characteristics and key functionalities of the access network and resources to users. It will simplify the access to the network resources, will give the opportunity for unification of the interfaces, will abstract the information related to the physical network resources and support simplified higher level interfaces for resource control.

The topology and user requirements awareness, and the possibility of self-organization and self-configuration, will ensure the necessary flexibility of the network, the possibility of resource and user discovery and the adaptation to the changes of users requirements, changes in the networks status and policies of the virtual resources owners. This awareness is essential for implementing intelligence of the access and the network, integrated management and operation, (including operations such as monitoring, fault detection, topology awareness, reconfiguration, resource scheduling and allocation, customized control, etc.), as well as network reliability and sustainability.

Mobility in the virtual access is related not only to the support of mobile users and services, but also movement of virtual resources such as computing resources, virtual access points and applications. Each virtual resource can be moved according to the users' demands and users can be dynamically attached or reattached to some virtual access point depending on the application characteristics. At the same time, to maintain the services continuity for the users, the services also can be moved together with the users without service downtime. In addition, the virtual resources can be added to improve network performance or removed for load balancing or energy saving purpose.

One major issue in the virtualization of the access will be the virtualization of the wireless links. Establishment of a two way wireless connection requires the control of wireless channel parameters such as power (uplink and downlink), operation frequency (including frequency reuse), receiver sensitivity, interference, etc., things that are quite different from optical fiber connections.

3.4.2 The Unified Virtual Network and Its Singularity

The access network beyond NGA should have all the characteristics of a Future Network which is able to provide revolutionary services, capabilities, and facilities that are hard to provide using the existing network technologies. Unlike the original IP based infrastructures focused on technical connectivity, routing, and naming, the scope of the future networks should encompass all levels of interfaces for services, manageability and technical resources (networking, computation, storage, control). Besides resource virtualization other characteristics of Future Networks are [53]:

- the accommodation of a wide variety of traffic and support diversified services (Service Diversity);
- flexibility to support and sustain new services derived from future user demands (Functional Flexibility);
- mechanisms for retrieving data in a timely manner regardless of its location (Data Access);
- having device, system, and network level technologies to improve power efficiency and to satisfy customer's requests with minimum traffic (Green Energy);
- facilitate and accelerate provision of convergent facilities in differing areas such as towns or the countryside, developed or developing countries (Service Universalization);
- be able to operate, maintain and provision efficiently the increasing number of services and entities (Network Management);
- to provide mobility that facilitates high levels of reliability, availability and quality of service in an environment where a huge number of nodes can dynamically move across the heterogeneous networks (Mobility);
- provide sufficient performance by optimizing capacity of network equipment based on service requirement and user demand (Optimization);
- provide a new identification structure that can effectively support mobility and data access in a scalable manner (Identification);
- support extremely high reliability Services (Reliability and Security).

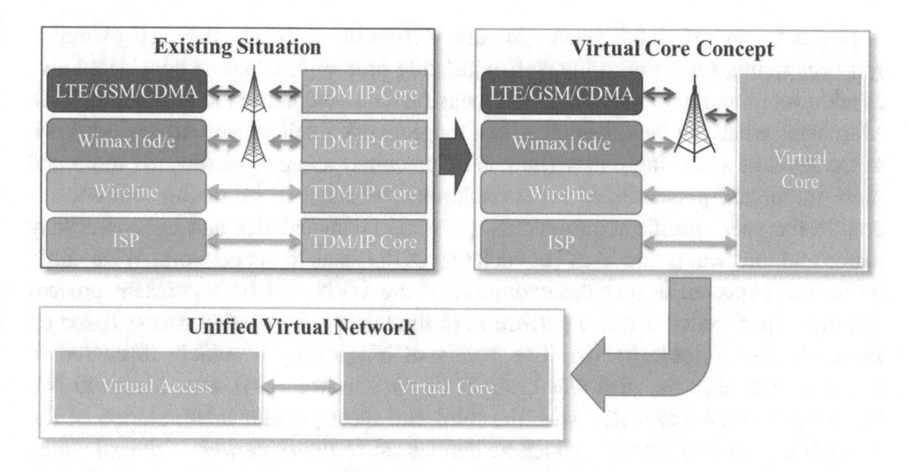

Fig. 3.6 Towards a unified virtual network

But what should be reasoned to be way beyond the NGA, is that the access network will become part of a Unified Virtual Network (UVN), as illustrated in Fig. 3.6. Such an UVN will have a unique and unified access infrastructure, based on Unified Virtual APs for wired and wireless access, organized through clear rules for hierarchy, territory, frequency allocation. This infrastructure will be accessed via one type of Unified Virtualized Interface (UVI) for all types of terminal equipment (computers, tablets, smart phones, machines, home appliances, industrial controllers, identification tags, etc.). This UVI interface will allow the implementation of novel access methods enabling migration to multi-access domains and permitting user subscriptions to different virtual access sub-domains. This will be a shared-facilities based access network with shared virtual access infrastructures and shared virtual cores, based on common approaches for virtualization of technologies and services, fully IP based and with cross-layered access "virtualizing" all different networks and substructures (operators, private networks, etc.). An automated access depending on location, user activity and required QoS, will be ensured through the self-planning, self-organizing, self-monitoring, self-regulating properties of the UVN. The UVN will be flexible, cognitive and with embedded and enabled intelligence in order to ensure dynamical development and evolution.

Regarding the future UVN development and evolution, the basic questions are: what will be the limits and in which way will this network evolve, and is there a way to envision its future evolution in the very long-term? The Shannon limit of course is the one that could bring the answer at first thought, as it gives the maximum possible throughput to be achieved in a communication link, or the so called "throughput wall". But as we can see even today, the addition of new dimensions in the network connections could provide practically unlimited throughput. Simple examples in the fiber optic and the wireless access are the technologies WDM and MIMO, which could be seen as the way to increase capacity by the addition of an additional spatial dimension in the connection. The implementation of intelligence in the network, such as the self-planning, self-organizing, self-monitoring, self-regulating properties could be considered also as addition of another technological dimension for overcoming the throughput wall. Moreover, what can be expected is that the network intelligence will spread down to the level of each single user (human or machine) and evolve towards more and more intelligent personalized access solutions, which will be taking into consideration the user specific requirements, personal characteristics and even everyday habits and the "wants and likes" of each one of the users in the network. What could be further expected is that the evolution of the UVN will be a constant process towards superintelligence and perfection of the telecommunication access based on *Technological Singularity* [8]. The notion of *Singularity* is widely discussed in future studies after the idea was brought up by Kurzweil [54]. Ray Kurzweil is a legendary futurist associated with the third and most popular interpretation of the "*Technological Singularity*", which he defines as: "*a future period … during which the pace of technological change will be so rapid, its impact so deep, that human life will be irreversibly transformed. Although neither utopian nor dystopian, this epoch will transform the concepts that we rely on to give meaning to our lives, from our business models to the cycle of human life, including death itself.*" Ray Kurzweil predicts, based on mathematical calculations of exponential technological development, that the Singularity will come to pass by 2045 and that in a post-Singularity world, humans would typically live much of the time in virtual reality, which would be virtually indistinguishable from normal reality.

All this comes to say that *Technological Singularity* is strongly associated with the development of "beyond human intelligence", based on computers that are superhumanly intelligent. In this relation, a very simple expectation is that the UVN with all its users (humans and machines) will become a *superhumanly intelligent entity*. It is also argued that beyond-human intelligence associated with *Technological Singularity* will spiral around a positive "*Singularity Feedback Loop*", where increased intelligence creates more and more powerful technology. Even today the world of computing has already cycled through a great many iterations in similar feedback loop [55]. With *Technological Singularity*, in such a spiral loop could enter the future *world-wide virtualized and unified communication network*. Finally, it is worth mentioning what John von Neumann's definition of singularity "*Singularity is the moment beyond which technological progress will become incomprehensively rapid and complicated*".

3.5 Conclusion

In this chapter the current access technologies and their limitations, together with today's view of what NGA is expected to be and its societal role, were presented. Issues such as advances and evolution of access networks, Fi-Wi convergence, current trends towards NGA and its development, were discussed. The expectations from the future access architectures, in order to meet the future and the challenges of a hyper-connected world, were considered. Based on the review and analysis of these issues, was brought up and presented the idea of the future core and access network development towards a Unified Virtual Network having unique and unified access infrastructure, based on Unified Virtual Access Points for wired and wireless access. The expectations behind the idea of such a network are that this will be a shared-facilities based access network with shared virtual access infrastructures and shared virtual cores, based on common approaches for the virtualization of technologies and services, "virtualizing" all different networks and substructures. This unified network will be organized through clear rules for hierarchy, territory and frequency allocation, will be a flexible and cognitive network with embedded and enabled intelligence and will ensure dynamical development and evolution. Finally, a vision regarding the future development and evolution of such a virtual access network is presented, and the way that the physical and theoretical limits of the throughput will be reached and even in a way exceeded. It is reasoned that the addition of new dimensions in the access network communication links, as well as the implementation of intelligence in the network that will spread down to the level of each single user, is the path that in practice will overcome the theoretical maximum communication link throughput. The future development and evolution will be a constant process towards superintelligence and perfection based on the notion of Technological Singularity and the access network with all its users (humans and machines) will become a superhumanly intelligent entity.

References

1. Thomson Reuters (2014) The world in 2025: 10 predictions of innovation. http://sciencewatch.com/sites/sw/files/m/pdf/World-2025.pdf
2. Cisco (2015) Cisco visual networking index: forecast and methodology, 2014–2019. http://www.cisco.com/c/en/us/solutions/collateral/service-provider/ip-ngn-ip-next-generation-network/white_paper_c11-481360.pdf. Accessed 27 May 2015
3. Tsagklas T, Pavlidou FN (2011) A survey on radio-and-fiber FiWi network architectures. Cyber J Multidisciplinary J Sci Technol, J Sel Areas Telecommun (JSAT) March Edition:18–24
4. Maier M (2014) Fiber-wireless (FiWi) broadband access networks in an age of convergence: past, present, and future. Adv Opt Article ID 945364. http://dx.doi.org/10.1155/2014/945364
5. Aurzada F, Lévesque M, Maier M, Reisslein M (2014) FiWi access networks based on next-generation PON and Gigabit-class WLAN technologies: a capacity and delay analysis. IEEE/ACM Trans Networking 22(4):1176–1189
6. Ali AM, Ellinas G, Erkan H, Hadjiantonis A, Dorsinville R (2010) On the vision of complete fixed-mobile convergence. J Lightwave Technol 28(16):2343–2357
7. European Commission (2014) Digital agenda for Europe. http://europa.eu/pol/pdf/flipbook/en/digital_agenda_en.pdf
8. Goertzel B, Goertzel T (eds) (2015) The end of the beginning: life, society and economy on the brink of the singularity kindle edition. Humanity+Press
9. Prasad R (2012) Future networks and technologies supporting innovative communications. In: Proceedings of the 3rd IEEE international conference on network infrastructure and digital content (IC-NIDC), Beijing, 21–23 Sept 2012. doi:10.1109/ICNIDC.2012.6418846

10. The Economist Intelligence Unit (2014) The hyperconnected economy: how the growing interconnectedness of society is changing the landscape for business. http://go.sap.com/docs/download/2015/10/04e64342-457c-0010-82c7-eda71af511fa.pdf
11. Vermesan O, Friess P (eds) (2015) Building the hyperconnected society—IoT research and innovation value chains, ecosystems and markets. River Publishers, Aalborg
12. Hamalainen S, Sanneck H, Sartori C (eds) (2011) LTE self-organising networks (SON): network management automation for operational efficiency. Wiley
13. GPP (2015) Technical specification 32.500 "telecommunication management; self-organizing networks (SON); concepts and requirements" Release 13
14. Sallent O, Pérez-Romero J, Sánchez-González J, Agustí R, Díaz-Guerra M, Henche D, Paul D (2011) A roadmap from UMTS optimization to LTE self-optimization. IEEE Commun Mag 49(6):172–182
15. Peng M, Liang D, Wei Y, Li J, Chen HH (2013) Self-configuration and self-optimization in LTE-advanced heterogeneous networks. IEEE Commun Mag 51(5):36–45
16. Andrade M et al (2011) Evaluating strategies for evolution of passive optical networks. IEEE Commun Mag 49(7):176–184
17. Effenberger F et al (2007) An introduction to PON technologies. IEEE Commun Mag 45(3): S17–S25
18. Kramer G (2005) Ethernet passive optical networks. McGraw-Hill
19. Effenberger F et al (2009) Next-generation PON—part II: candidate systems for next-generation PON. IEEE Commun Mag 47(11):50–57
20. Zhang J et al (2009) Next-generation PONs: a performance investigation of candidate architectures for next-generation access stage 1. IEEE Commun Mag 47(8):49–57
21. Kani JI et al (2009) Next generation PON—part 1: technology roadmap and general requirements. IEEE Commun Mag 47(11):43–49
22. ITU-T (2013) G.987: 10-Gigabit-capable passive optical network (XG-PON) systems: Definitions, abbreviations and acronyms
23. Hajduczenia M, da Silva H, Monteiro P (2007) 10G EPON development process. In: Proceedings of the 9th international conference on transparent optical networks, vol 1, Rome, 1–5 July 2007. doi:10.1109/ICTON.2007.4296087
24. Park SJ et al (2004) Fiber-to-the-home services based on wavelength-division-multiplexing passive optical network. IEEE J Lightwave Technol 22(11):2582–2591
25. Wagner SS, Kobrinski H (1989) WDM applications in broadband telecommunication networks. IEEE Commun Mag 27(3):22–30
26. Effenberger F, Lin H (2009) Backward compatible coexistence of PON systems. In: Proceedings of the conference on optical fiber communication, San Diego, CA, 22–26 March 2009
27. McCammon K, Wong S W (2007) Experimental validation of an access evolution strategy: smooth FTTP service migration path. In: Proceedings of the optical fiber communication and the national fiber optic engineers conference, Anaheim, 25–29 March 2007
28. Choi K et al (2007) An efficient evolution method from TDM-PON to next-generation PON. IEEE Photonics Technol Lett 19(9):647–649
29. Chen J et al (2010) Cost vs. reliability performance study of fiber access network architectures. IEEE Commun Mag 48(2):56–65
30. Kazovsky L et al (2007) Next-generation optical access network. IEEE J Lightwave Technol 25(11):3428–3442
31. Shami A, Maier M, Assi C (eds) (2009) Broadband access networks, technologies and deployments. Springer
32. Kitayama K, Wang X, Wada N (2006) OCDMA over WDM PON-solution path to gigabit symmetric FTTH. IEEE J Lightwave Technol 24(4):1654–1662
33. Fabrega JM, Vilabru L, Prat J (2008) Experimental demonstration of heterodyne phase-locked loop for optical homodyne PSK receivers in PONs. In: Proceedings of the 10th anniversary international conference on transparent optical networks, vol 1, pp 222–225, Athens, Greece, 22-26 June 2008

34. Cvijetic N (2012) OFDM for next-generation optical access networks. IEEE J Lightwave Technol 30(4):384–398
35. Charbonnier B, Brochier N, Chanclou P (2011) (O)FDMA PON over a legacy 30dB ODN. In: Proceedings of the optical fiber communication conference and exposition, Los Angeles, 6–10 March 2011
36. Wong E (2012) Next-generation broadband access networks and technologies. J Lightwave Technol 30(4):597–608
37. Sarigiannidis AG, Iloridou M, Nicopolitidis P, Papadimitriou G, Pavlidou FN, Sarigiannidis PG, Louta MD, Vitsas V (2015) Architectures and bandwidth allocation schemes for hybrid wireless-optical networks. IEEE Commun Surv Tutorials 17(1):427–468
38. Raaf B et al (2011) Vision for beyond 4G broadband radio systems. In: Proceedings of the 22nd IEEE international symposium on personal, indoor and mobile radio communications, Toronto, Canada, pp 2369–2373, 11–14 Sept 2011
39. Andrews JG, Buzzi S, Choi W, Hanly SV, Lozano A, Soong ACK, Zhang JC (2014) What will 5G be? IEEE J Sel Areas Commun 32(6):1065–1082
40. Boccardi F, Heath RW, Lozano A, Marzetta TL, Popovski P (2014) Five disruptive technology directions for 5G. IEEE Commun Mag 52(2):74–80
41. EU Horizon 2020 (2015) Future of 5G networks. http://ec.europa.eu/digital-agenda/en/towards-5g
42. Maier M (2014) Fiber-wireless (FiWi) broadband access networks in an age of convergence: past, present, and future. Adv Opt Article ID 945364. http://dx.doi.org/10.1155/2014/945364
43. Ghazisaidi N, Maier M (2011) Fiber-wireless (FiWi) access networks: challenges and opportunities. IEEE Netw 25(1):36–42
44. China Mobile Research Institute (2014) C-RAN white paper: the road towards Green Ran. http://labs.chinamobile.com/cran
45. Checko A, Christiansen HL, Yan Ying, Scolari L, Kardaras G, Berger MS, Dittmann L (2015) Cloud RAN for mobile networks—a technology overview. IEEE Commun Surv Tutorials 17(1):405–426
46. Wu J, Zhang Z, Hong Yu, Wen Yonggang (2015) Cloud radio access network (C-RAN): a primer. IEEE Netw 29(1):35–41
47. Chih-Lin I, Huang J, Duan R, Chunfeng C, Jesse (Xiaogen) J, Lei L (2014) Recent progress on C-RAN centralization and cloudification. IEEE Access 2:1030–1039
48. ITU-T (2012) Recommendation Y.3011 framework of network virtualization for future networks
49. ITU-T (2014) Recommendation Y.3012 requirements of network virtualization for future networks
50. Baumgartner A, Reddy VS, Bauschert T (2015) Mobile core network virtualization: a model for combined virtual core network function placement and topology optimization. In: Proceedings of the 1st IEEE conference on network softwarization, London, 13–17 April 2015
51. Liang C, Yu FR (2015) Wireless network virtualization: a survey, some research issues and challenges. IEEE Commun Surv Tutorials 17(1):358–380
52. ITU-T Recommendation Y.3011 (2012) Framework of network virtualization for future networks
53. ITU-T Recommendation Y.3001 (2011) Future networks: objectives and design goals
54. Kurzweil R (2005) The singularity is near: when humans transcend biology. Viking
55. Barnatt C (2016) A guide to the future. http://www.explainingthefuture.com/index.html

Chapter 4
Ubiquitous Wireless Communications Beyond 2050

Shu Kato

Abstract This chapter looks back our wireless communications R&D history and speculates what will happen "35 years later", 2050. This intends to show the importance of foreseeing the Real Goal (behind the goal) for coming generations of cellular phone networks in addition to the importance of integration of cellular/mobile spectrum and WLAN/WPAN networks. Furthermore, to be successful in this area, it is important to try "What to do approach", and "application driven technology development" instead of technology oriented R&Ds.

We started the concept of FPLMTS (Future Public Land Mobile Telecommunication Systems) to enable to communicate with anybody, anywhere and anytime at a 1.5 Mbps transmission rate on move about 30 years ago. This demanded the common spectrum to be able to communicate globally assigned that turned out to be impossible by the fact "having auctioned the spectrum" by some countries. On the other hand, technology advancement made it unnecessary—the purpose of common spectrum has been achieved by multi-bands RF, automatic network finding and so on effectively with very minor cost up. Currently we are happy to communicate with anybody, anywhere and anytime with one handset. The real goal of the FPLMTS has been met in a different way that we need to notice as an important example not to seek the goal by "one way".

S. Kato (✉)
Tohoku University, Sendai, Japan
e-mail: shukato@riec.tohoku.ac.jp

R. Prasad and S. Dixit (eds.), *Wireless World in 2050 and Beyond: A Window into the Future!*, Springer Series in Wireless Technology,
DOI 10.1007/978-3-319-42141-4_4

4.1 Hard Time for Wireless Communications in Late 1980s—About 25 Years Ago—We Live in Very Lucky Time Now

There came out an article titled "Radio Is Dead" on Communications Magazine by IEEE COMSOC. Then President-elect came to Satellite and Space Communications Committee and suggested to get merged with Radio Communications Committee or get lost. Yes, long haul radios were dead but not mobile and WLAN/WPAN systems.

At those times in COMSOC Technical Committees, only two were related to wireless: Radio Communications (RC) Committee, and Satellite and Space Communications (SSC) Committee among some 20 TCs. Now, there are more than 20 Technical Committees related to "wireless" except one or two "optical fibers" related TCs. This is a huge change of the wireless researchers' status in academia as well as in industry.

4.2 When Digital Cellular Started—Competition Made Cell Phones Commodities (20 Years Ago)—No More Carrier-Made Market

Until then, carriers' logic decided networks and services since it was the only available network and the customers' choice was "to use or not to use" it. Then, multiple carriers started to offer "similar" services even if with the same air interface.
Then, the competition war started over price, coverage and available applications. Paradigm shift started in mid 1990s from carrier to customer centric communications services. Some major changes occurred, such as

i. From technology oriented system design to service/applications oriented system design,
ii. Capacity was an issue but "business success—customers' satisfaction" was more important.

A good competition example was seen in Jacksonville, FL (Residents: 0.5 M) in late 1990s. With new entrants prices dropped 46 % in a matter of months: Customers selected a carrier based on best prices, features and services (Table 4.1).

- Quality
- Price
- Features

Table 4.1 Seven competitors in the market in Jacksonville

BellSouth	Prime Co.
AT&T	Nextel
Powertel	Alltel
Sprint	

4.3 Mobile Phones Become Commodities Now

This means that (i) customers care for Price, Quality, Features but not Technologies at all, and (ii) Students are not much interested in cell phone technologies any more.

More than 4.5 billion mobile users among of 7.0 billion people on the earth exist and they have created necessity of more spectra for mobile communications and created necessity of lower power consumption/Green communications as well.

People demand more capacity and better quality in audio and video transmissions. Although we are missing young students' interests in cell phones due to "too commodities" to them, they demand more capacity for gathering information and dissemination of information. They demand higher quality audio/video transmission—"music CD" level audio codecs will come to mobile market pretty soon and uncompressed HD video will come to market soon too. The latter may need more bandwidth, a couple of GHz/channel that mandates the use of 60 GHz or higher frequency bands [1–3].

4.4 Lessons Learned—Failed Too Technological Approach

Let's review what happened on voice codec for mobile applications. The analog mobile networks such as AMPS (Advanced Mobile Phone Networks) necessitated 30 kHz/voice transmission and hexagonal frequency reuse structure required 210 kHz/voice transmission 2-dimensionally. A lot of competitive developments were going on in 1990s towards lower bit rate codecs so that more voice channels can be transmitted over the analog voice channel bandwidth, 30 or 210 kHz. Here we will see three different voice codecs:

1. American cell phone codec
   American digital cell phone started with VCELP (Vector Code Excited Linear Prediction: a kind of CELP (Code Excited Linear Predictive) codecs) and first attempt with the standard IS-54 ended up with a failure due to voice quality. Although this codec realized three times spectrum efficiency against analog cell phone systems, the voice quality was not there—carrier's logic was there but no

customer satisfaction. This codec extracted voice signal specific spectrum and the coded information of the spectrum is sent to the receiver (but not sending real voice signals) and the receiver will create the same spectrum to drive the speaker locally. This caused "mechanically sounding" voice signals. This was good to compress voice signals to accommodate more voice channels over the limited bandwidth (carrier's logic); however, this was not accepted by customers. America stopped this IS-54 services and later came back to digital cellular market with much better voice codec, IS-136 in late 1990s. NTT Docomo used almost the same codec with IS-54 codec and was not popularly accepted in Japan either.

2. Japanese PHS
   PHS (Personal Handy Phone Systems) was developed in Japan in early 1990s with a good quality codec, ADPCM. Japanese PHS system was developed to replace indoor cordless telephones and this is why ADPCM codec was adopted as its quality is supposed to be as good as landline telephone networks with PCM codecs. This was well accepted with much better voice transmission quality than cellular phones in Japan although it had a mobility issue due to its inherent characteristics—targeting indoor use and allowing outdoor use as an incentive. There was no way for PHS phones to compete with cell phones in moving environments that was not the one PHS is designed for.
3. GSM
   GSM was standardized much earlier than American or Japanese cellular phone systems and luckily enough there was no low rate voice codec technology available. GSM adopted 13 kbps voice codec that is about twice the bandwidth of VCELP in IS-54. This wider bandwidth allocated for voice transmission, and low technology voice codec helped GSM to keep voice transmission quality high and GSM survived as relatively good voice quality cellular phones globally.

4.5 What We Should Do to Satisfy Customers 35 Years Later?

It is important to set the goals first, such as we need more capacity, better coverage, better quality and better applications with lower power consumption. How can we achieve them? We should be flexible in thinking and seek multiple approaches to hit the goal including interference cancellation, fading mitigation, higher modulations, stronger FECs, etc. The way of thinking should be "what should be achieved" but not "how it should be done" (Fig. 4.1). We should not stick to "Mobile approaches" but need to look at the goal by much wider ways—WLANs and WPANs may compliment narrow band mobile access networks to level off peak traffic very well [4–6].

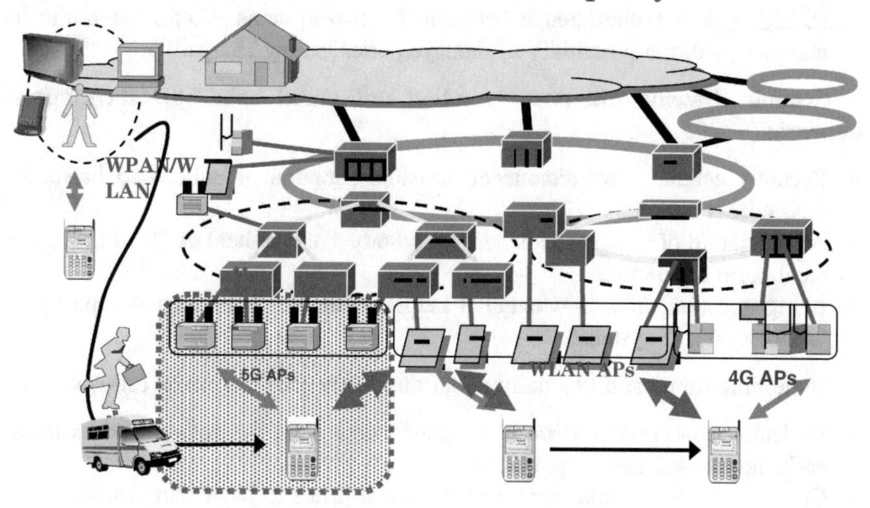

Fig. 4.1 WLAN/WPAN in future mobile communications Era

As of 2016, over 75 % of data traffic has been carried by WiFi networks and cellular networks are no more major players in this sense. A heavy discussion has been going on collaboration between WiFi and cellular guys for 5G (Fifth Generation) cellular networks access recently [7]. This trend will continue to grow for coming years due to mobile spectrum shortage. To meet this demand, there have been a lot of activities in the IEEE 802 LAN/MAN Standard Committee on

i. increasing transmission capacity/Hz in existing bands (60 GHz)
ii. developing a new spectrum for wider bandwidth such as Teraherz.

By adopting WPAN/WLAN systems for cellular access networks, we can save scarce mobile spectrum for the terminals on move and provide wider capacity for static terminals.

- Not everybody needs MOBILITY.
- Spectrum for mobile communications: Scarce.

4.6 Integrated Cellular and WiFi

- Two distinct approaches are being taken now: (i) Use WiFi to level off the peak traffic- on going, and (ii) Cellular systems to use unlicensed bands currently used for mainly WiFi—with licensed band control channels.

- With approach (ii), there is a need for harmonized Clear Channel Assessment (CAA) to avoid interference between the two systems. Could the traffic be charged (with the possibility of damaged after getting charged)?

Cellular Systems will possibly adopt unlicensed bands for their access networks

- Recently cellular guys announced possible adoption of unlicensed bands for their access networks
- As the nature of unlicensed bands, everybody can use them as far as they meet local radio regulations
- A big question mark is whether it helps to enhance transmission capacity of "cellular+WiFi" systems.

Some time-frame and key point about this likely scenario are as follows.

- Basically, co-existence issue is an independent issue for both parties as far as each meets local legal regulations
- Co-existence for cellular guys and WiFi is a "Nice to have" but not "must"
- IEEE802 can suggest how to test but cannot certify cellular products
- It looks like WiFi Alliance has not been testing the co-existence performance (obviously no test by IEEE802)
- The best way might be that IEEE802 inputs "how to test co-existence" to cellular guys and they implement them so that co-existence performance data could be shared by both parties.

4.7 Who Will Win 35 Years Later?—Applications Oriented Technology Development

Nobody cares about the network—assuming mobile networks become backbone as optical fiber networks are today for internet, the concern would be only about price, capacity, and applications.

What we need to do as engineers is dictated by what the consumers demand, such as applications including robust service coverage, enough bandwidth per customer, applications peculiar to mobile, and technologies that provide good coverage, high capacity and low cost. One essential problem, independent from applications is scarce spectrum to offer these functionalities. We may need some new technologies to double/multiple the current spectrum efficiency by 2050.

4.8 Conclusion

What but not how approach will open up more opportunities—A speculation on "2050 and beyond".

In the mobile world, cellular phones will become just commodities, carriers' profile will become lower and less important with time, and technologies to improve spectrum efficiency will be forever important.

The purpose of the technology development should be studied deeply as there are many alternatives to realize the real goal. Success would depend on what kinds of new applications are created and how quickly they are deployed, i.e., the technology development should be applications oriented. Finally, we may enjoy enough coverage and enough bandwidth over the mobile networks well assisted with WLAN/WPAN technologies in 2050.

References

1. Kato S, Harada H, Funada R, Baykas T, Sum CS, Wang J, Rahman MA (2009) Single carrier transmission for multi-gigabit 60-GHz WPAN systems. IEEE J Sel Areas Commun 27 (8):1466–1478
2. Wang J, Lan Z, Pyo CW, Baykas T, Sum CS, Rahman MA, Gao J, Funada R, Kojima F, Harada H, Kato S (2009) Beam codebook based beamforming protocol for multi-Gbps millimeter-wave WPAN systems. IEEE J Sel Areas Commun 27(8):1390–1399
3. Baykas T, Kato S et al (2011) IEEE 802.15.3c: the first IEEE wireless standard for data rates over 1 Gbps. IEEE Commun Mag 49(7):114–121
4. Kato S (2011) IEEE finds broader 4G wireless access will accelerate economic development. IEEE News Release
5. Sawada H, Takahashi S, Kato S (2013) Disconnection probability improvement by using artificial multi-reflectors for millimeter-wave indoor wireless communications. IEEE Trans Antennas Propag 61(4)
6. Kato S (2010) Deep future: beyond IMT-advanced and the need of additional spectrum. Panel session of PIMRC 2010, Istanbul, PIMRC2010 Panel
7. IEEE Standardization doc (2015) IEEE 802.19-15/0069r7

Chapter 5
The Networkless Network

Rajarshi Sanyal

Abstract The advancements in semiconductor technologies had a significant impact on the evolution of the mobile networks and the devices. With the advent of smartphones, it is now possible for telephony and VAS to be rendered by platforms which are not a part of the mobile network itself but powered by 'Over the Top' players like Whatsapp. The mobile network as it stands today, plays the role of delivering the fat pipe to the mobile users required for these applications. The network itself is lesser involved in delivering VAS to its users. Never the less, we still need a network to serve its mobile users. Is it feasible for a mobile device to communicate with another mobile device bypassing the mobile network for all services and agnostic to the distance between the devices? Staying within the framework of 5G, 'Device to Device' communication may be possible but only when the devices are in proximity with each other. It is imperative that we need to explore the feasibility to develop a solution where the devices can communicate and roam in absence of a layer 7 network and remain agnostic to the distance between the devices. If that is possible, we may find a way to replace the existing layer 7 mobile network with just a layer 1 access. The outcome is obvious: 'A Network-less Network'. The prosaic signalling procedures related to a mobile network, specifically addressing, mobility and location management can be directly actuated by the mobile devices. In this chapter we investigate the viability and practicality of developing and implementing such a network solution.

A mobile network facilitates its users to share a common network resource, i.e., spectrum, to place voice and video calls, to roam, send SMS, use packet data services within the network area and avail value added services using the communication channels like voice, SMS and data. The available spectrum is sliced into multiple radio frequencies and these radio frequencies are allocated to the hexagonal cells of a cellular network. The mobile network has switching capabilities

R. Sanyal (✉)
Belgacom International Carrier Services, Brussels, Belgium
e-mail: rajarshi.sanyal@bics.com

R. Prasad and S. Dixit (eds.), *Wireless World in 2050 and Beyond: A Window into the Future!*, Springer Series in Wireless Technology,
DOI 10.1007/978-3-319-42141-4_5

which enables it to make and break circuits between the users enabling them to originate or terminate calls. The same goes for packet data service where the mobile network needs to activate or delete the packet data context initiated by smartphones or by the network. For all these operations, a mobile network requires intelligent network elements which are used for mobility and location management of the mobile subscriber, cell to cell handover, switching of calls, control signalling, supplementary service (call conference, caller line identification, etc.) and managing data bearers per subscriber. The mobile network is heavily dependent on the network to maintain its location and presence within the network area. The handset invokes many control messages towards the network continuously (and vice versa) even in the idle mode to update the network on its location and presence. Throughout its evolution phase (2G to 5G), the mobile networks did not deviate from this philosophy implying that there is incessant control signalling exchanged between the device and the network for the network operations. This has some impacts. The network requires chunks of processing power and bandwidth to manage device location and maintain contact with millions of users. The mobile devices are heavily dependent on the network for these operations. The network architecture would require a major revamp to cater the pre-requisites of a new generation of services, like Device to Device communication, Internet of Things.

It is imperative that the fundamental notion behind of the present cellular network (5G) design may need to be reinvestigated. The foundation of a cellular network is the OSI (Open Systems Interconnection) which was coined 43 years back. Typically, the physical layer (the medium, like copper or fibre, air medium like microwave) carries the generic traffic data for the network users, but the processes that require network intelligence (like routing, mobility and location management) are actuated at higher layers like layer 3 (network layer), layer 5 (presentation layer) and application layer (application layer). The involvement of the higher layers of OSI increases the network processing and resources, overall intricacy, cost of network and the devices and overall power requirement.

The key questions here are:

- Can improvisations in the mobile device stacks, the addressing scheme and the modulation technique used for the communication, lead to mobile handsets which behave more like walkie-talkies depending only on the physical layer of the OSI, but still enjoy the mobility and roaming?
- Is it possible to realise a communication system without an intelligent radio network?
- Is it possible to have a mobile network with passive signal repeaters and channel aggregators, eliminating intelligent switching and mobility management elements, intelligent radio access networks (eNodeBs), etc.?

If we can find a practical approach to realise such a network, then this can be the blueprint to instil Device to Device communication that is agnostic to spatial bounds.

Let us investigate how this may be achieved. We take a simple example, this time an old generation radio calling systems for the cabs.

- There is no network here. Just a central transceiver at the Cab station and the radio sets fixed at the cars.
- It is typically a broadcast network. The cab chauffer gets continuous announcements, waits to pick up the name, number, location etc. for his turn.
- This information is processed by the chauffer's brain, and not by the network.
- If the cab company decided to channelize only the relevant information to the cab drivers and not to send a broadcast signal, then they would have needed some intelligence somewhere in the network to do this. Ideally this could have been realised by using mobile network (instead of a radio broadcast network) which switches the relevant information via a radio channel towards a mobile handset available with the chauffer now (instead of the non-intelligent radio transceiver as we discussed before).
- Let us call this intelligent entity sitting in the network as MR X, and let us call the chauffer MR Y.
- The present generation mobile network including 5G (comprising of the active network components in the radio and core layer) acts as MR X. So he has to sit somewhere within the network. Mr X's real analogue in a mobile network is: the radio and core network adjuncts (BTS/BSC/Mobile Switching, GPRS support Nodes etc.). Specifically in 4G, they are like the eNodeB, MME (Mobility Management Entity), SGW (Serving Gateway), PGW (Packet Gateway) etc.
- With the proposed methodology of conceiving a network-less network, the ONLY active network component acts as an analogue of MR Y sitting behind his radio trunking set in the car. There is no real intelligence required in the network anymore. The physical layer of the network along with the intelligence (MR Y) in the handset will be able to pick up the relevant information for it.
- Off course we still need the core network processes for Authentication, Authorisation, Accounting (Charging, billing), all the supplementary services (caller line identification, call conference, SMS, MMS) and VAS, legal intercepts etc.
- Such a network does not continuously need to ping the handset (and vice versa) for mobility management, channel control, cell to cell handover etc. as it happens in 4G networks.
- The network architecture and the network processes are simplified to a considerable extent.
- Device to Device communication at global scale will ensue.

Time has come to think beyond the conventional network principles to design a new breed of networks.

5.1 The 'as Is' Network Landscape

Mobile communication technologies have evolved over the years. From the onset of 2G and its metamorphosis towards 5G, the bandwidth per user had increased phenomenally. This has been endorsed by advancements in access technologies, core network technologies as well as electronics. Needless to mention that the society at large had changed over the years and the technology had evolved in line with the social requirements. People would like to stay connected with each other via peer to peer communication applications or social media applications. Most of us use the mobile device or the tablet to reach the social media platform. The devices are equipped today with the required processing power, memory and the sensors that does not only provide the leeway to avail such services but also makes it possible to render network services like location and presence without the involvement and support of the core mobile network. Thanks to VLSI and silicon technologies, the processing power, the volatile and the static memory available few years back in desktop is now available in the smartphones. But this has a knock on effect on the entire ecosystem. As the end devices become more powerful, the networks functionalities especially those related to VAS become more redundant. This has sparked off a million dollar debate. Who is presently at the control center? Is it the network operators or the bunch of developers continuously rendering new communication and social media 'Over the Top' applications that can run in those end user devices?

With the new communication technologies, spectral efficiency has increased manifold. Cost for connectivity and bandwidth has dwindled. Every operator aims to render the best coverage and throughput for its users at better rates than its competitor. Once you have that, you can use any applications on the smartphone for basic services, like voice and messaging and also advanced services like interactive gaming, 3D video chat etc. The mobile network is slowly but surely congregating towards a fat pipe approach. Services are offered more and more over the top and lesser by the network itself. In the process, the mobile operators are losing their proximity towards their subscriber and hence the reducing their share of the pie. If we need to use Facebook today, we will identify ourselves more with Facebook platform rather than the mobile operator who is providing me the data connectivity required for browsing the Facebook app.

Well, the operators decided to play the last ditch battle. GSM association came up with a network centric technology for rich messaging alike Whatsapp or Viber which is typically referred as 'Over the Top' services. This network driven rich messaging application is called Rich Communication Suite (RCS). RCS, which is more familiar with the name Joyn (a trademark for RCS) is essentially a thread of intricate standards and protocols. The mobile network needs to deploy a new core called IMS (IP Multimedia System) on top of existing LTE core just for this messaging service. RCS did not take off yet. But as new voice driven technologies like Voice Over LTE (VoLTE) are in the process of deployment which necessitates the IMS core on top of 4G network, RCS may naturally evolve on top of VoLTE as the default rich messaging app. This also has been rekindled by the news of Google's recent takeover of Jibe who is a RCS supplier.

So it is imperative that the trend is changing a lot. A decade back, value added services were rendered by the mobile network. Protocols like Parlay, messaging over IP, location based services, and presence were used by Mobile Service Delivery platform to render these specialised and orchestrated services. Over the years the significance of these VAS network elements have reduced as most of these service can be delivered by the OTT applications with virtually no dependence on the network.

We cannot still say that the OTT applications have made the mobile networks redundant and obsolete. A mobile network is still essential to deliver mobile service not only because it provides the connectivity but also manages mobility, user location and addressing.

However the major OTT players like Google and Facebook are churning out millions of dollars to figure out ways to diminish the dependence of their users on mobile networks.

5.1.1 Moving Towards Free Internet: Google Loon, Facebook Drone and Cubic Satellite Projects

Google project Loon is a network of balloons travelling approximately 12 miles above the Earth's surface in the stratosphere, designed to free internet to the people. Project Loon uses software algorithms to determine where its balloons need to go. Sailing with the wind, the balloons can be arranged to form one large communications network.

The network to be realised via Facebook's drones will create a network for free internet similar to Google's Project Loon. While Loon uses balloons instead of drones, the airborne elements in both networks distribute signals to each other to increase range.

Lasers will be used for the drones to communicate with each other, while the drones will communicate with the ground using radio signals. "A ground station will transmit a radio Internet signal to a mother aircraft that will then feed other aircraft in the constellation using laser technology," with the drones sending radio Internet signals down to users on the ground, Facebook explained

A group of individuals in New York are trying to give the world free global internet via a network of miniature satellites broadcasting Wi-Fi down from space.

Outernet, has been working since December 2013 to aims to provide one-way internet, which is essentially a 'downlink only' internet connectivity to all citizens. The typical use case may be broadcasting information like weather, warnings, crop related information, stock information, news etc.

Outernet's Wi-Fi solution works by using hundreds of tiny 10 cm cube-shaped satellites called "cubesats", which are cheap to produce. The constellation of cubesats would use standardised radio protocols as well as WiFi multicasting.

5.1.2 *The Network for Machines: Lora—Low Power—Long Range*

The network requirements for machine to machine communications had triggered the development of a new breed of networks termed as 'Low Range' or Lora. In a traditional cellular network, there are some constrains in catering the M2M applications. When the cellular network was designed initially, it was meant to serve only the human user. With the advent of 'Machine to Machine' devices and 'Internet of Things', the traditional networks faced a major problem as the behaviour of this new generation of devices and their traffic characteristics are markedly different. Profiles of millions of devices need to be provisioned in the network which may subsequently generate only meagre volume of traffic due to nature of the application (say electric metering devices) but still remain attached to the network continuously. If these devices are in motion, the network also needs to steer the mobility management process. The machines can generate flash crowd traffic. All these factors can jeopardise the performance of the overall network hence impacting the quality of service rendered to human users too. This is the reason why we witness some initiatives to build networks dedicated for M2M communications. These networks tend to abandon the traditional cellular network design and move towards a bit-pipe approach in line with the principles followed by physical layer addressing. One such promising technology is Lora. LoRa is a wireless technology that has been developed to enable low data rate communications to be made over long distances by sensors and actuators for M2M and Internet of Things. It is a 2-way wireless solution that complements M2M cellular infrastructure, and provides a low-cost way to connect battery operated and mobile devices to the network infrastructure. A Lora device gets attached to LoRa concentrator gateway. The addressing scheme is unique utilising non-E.164 or IP parameters which provide the leeway to cater millions of devices with ease. Lora also improves the battery lifetime of your end-user devices, while minimizing signal interference and implementing a unique scheduling algorithm.

5.2 'To Be' Communication Landscape: Moving Towards Device to Device Communication with no Bounds

Our primary objective is to propose a technique of mobile communication where we render intelligence to the physical layer (of OSI) to take part in some processes which is otherwise confined to higher layer signalling activities, like, for example, addressing of a node. If we have a method to actuate user identification for the purpose of mobility management by implementing lower layer processes, we can simplify or eliminate the layer 7 (of OSI) processes and hence reduce the intricacy and costs of mobile network elements and user equipment.

The smart modulation scheme realised via SMNAT) abets the identification of the network users right at the physical layer and relieves the application layer from the prosaic activities related to addressing and mobility to a considerable extent. The notion of the physical layer addressing is realised by the dual concentric ring modulation scheme (Fig. 5.3) where the data symbols are not only meant for carrying the user information like traditional multiple access schemes in purview of GSM/CDMA/LTE. Instead they are also used to depict the user right at the physical layer. This is a way to realise 'addressing' using physical layer techniques. All mobility related processes like handover can be directly processed by the physical layer, thanks to the unique checkerboard cell topology (addressed in details in my paper).

Noticeably, in all existing mobile network topologies, modulation schemes are designed solely for the purpose of conveying data symbols for all users in the complex plane (user identification is not possible in layer 1). We propose to drift from this paradigm and actuate user identification and mobility at physical layer. As a direct consequence, the incessant signalling inter-exchange between the network and the devices reduces manifold, as the network does not need to continuously track the mobile station within the coverage area.

Figure 5.1 explains this process at high level. An existing mobile network node, be it a network element (like Mobile Switching Center) or a handset needs to analyse the transceived data at network layer and application layer of the OSI to identify the user. With our approach, this can be achieved right at the physical layer,

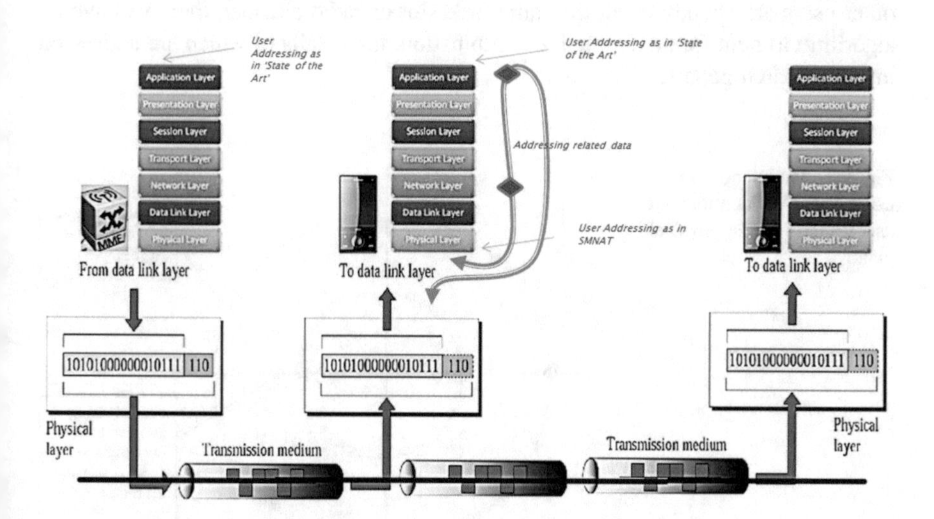

Fig. 5.1 User addressing in present day networks compared to our proposed scheme based on SMNAT

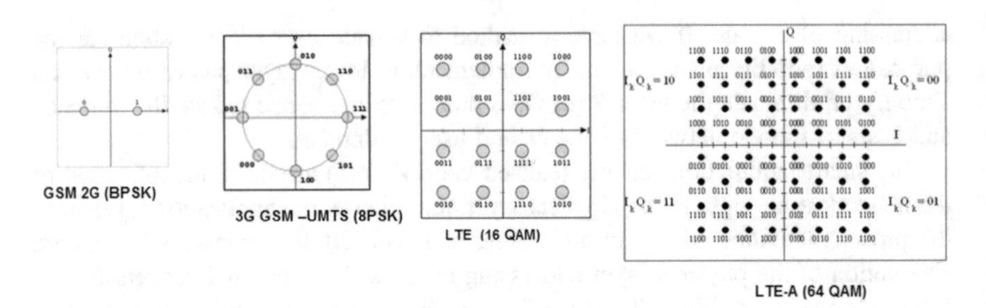

Fig. 5.2 Conveyance of data symbols for all users in the complex plane w.r.t the mobile network modulation schemes

i.e., the complex plane which carries the data symbols for all users pertaining to a particular modulation scheme.

In Fig. 5.2 this aspect has been elaborated further.

Following the evolution trail of a mobile network from 2G to LTE-Advanced (5G) as in Fig. 5.2, we do not discern a major philosophical shift in how user addressing and mobility management is achieved. The data symbols as you can see in the complex plane are meant for all users. As explained in Fig. 5.1, the existing network has to spawn intricate processes to meet these objectives.

However with SMNAT, user addressing can be done directly at complex plane. In Fig. 5.3 below, we demonstrate as a Proof of Concept that we can realise two concentric rings as the modulation scheme. In the outer ring, we assign a unique symbol with fixed coordinate in the complex plane for each user. In the inner ring, we have data symbols for conveying user traffic. Solely for the purpose of addressing, the user is also assigned a specific timeslot and channel (Fig. 5.4) If the other users are already using the same time slot or radio channel, then we have an algorithm to hunt for an alternative combination, the details of which are addressed in my research papers.

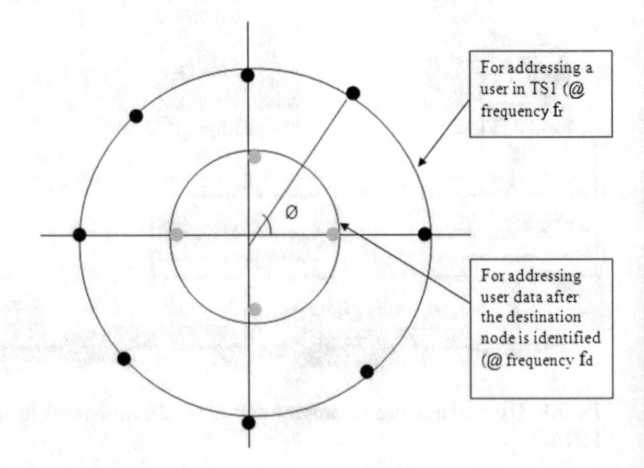

Fig. 5.3 Modulation scheme using SMNAT identifies the user directly in the physical plane

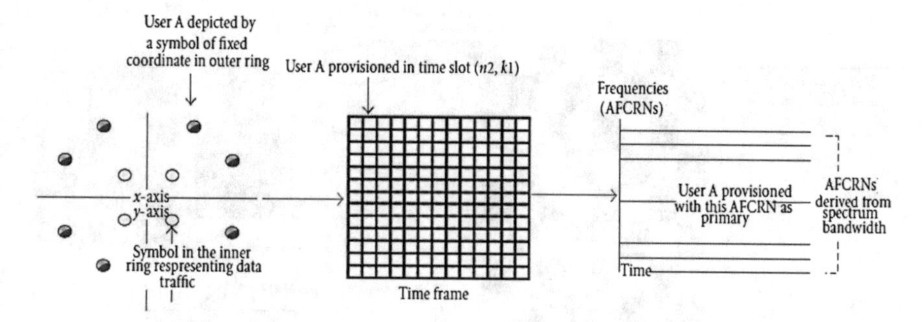

Fig. 5.4 Symbol coordinate, time slot and radio channel makes a user unique in SMNAT

5.2.1 Handovers

When a dynamic mobile user is in conversation state and moving from one cell to the other, the network should ensure that the call does not drop. In early stages of cellular network, we had hard handoff, where the circuit was broken temporarily and then re-established in the new cell. The cut in speech was momentary and unnoticeable. This method was typically referred to as 'break before make'. In newer generation of cellular networks, we have soft handoff where there is no break in the circuit, often called 'make before break'. In either of these two processes, the network and the mobile device need to exchange control signalling information between them to ensure that the call continues smoothly in the new cell. Needless to mention that both these processes are network driven and network controlled.

With SMNAT, the handover process is significantly different. The process is device driven rather than network driven. Moreover it is a physical layer process and does not require knotty signalling data exchange between the device and the network. In Fig. 5.5, the handover process is shown, which is self-explanatory. The whole network area is divided in square shaped cells and not hexagonal cells like existing networks. Frequency allocation is simple and no frequency planning needs to be done whereas we at present need to assign specific frequencies to specific cells and avoid frequency reuse in neighbouring cells. In SMNAT, we have only 2 kinds of cells, black and white where a fixed block of frequencies are allocated and this process repeated over and over again for all cells in the network area. We have a dual radio handset, one radio clamps on to the active cell (pertaining to white or black depending upon the signal strength). This is the active radio. The passive radio listens to the channel conditions pertaining to the nearest cells of the opposite colour. The channel transition during handover from the old to the new cell happens internally in the dual radio device from the previously active radio to the 'now' active radio.

As the interaction between with the network for the handover is essentially a physical layer process, we envisage a significant drop in the number of the

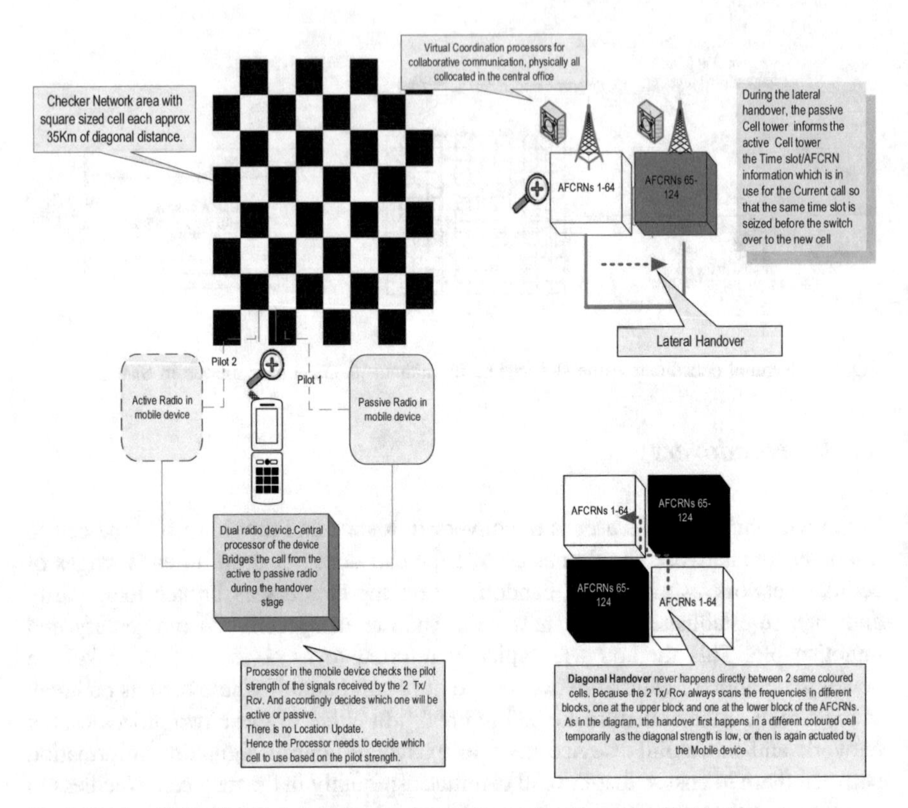

Fig. 5.5 The dual radio device driven handover process based on physical layer

intelligent processes in the network that need to be spawned to actuate a handover when compared to the 'State of the Art' mobile network.

What are the benefits?

1. Moderate control signalling, implying savings on bandwidth, network resources
2. Handover for device to device communication can be made very simple as we are reducing the interaction with network.

5.3 Application Scenarios for Device to Device Communication Realised by SMNAT

Nano mobile devices designed for 'Human to Human' or 'Machine to Machine' applications have constrains to operate in a mobile network, due to limited power resource and restrictions in RF power output. The present generation mobile network proposes several approaches, for example, deployment of femto cells, implementation of LTE direct between the mesh of devices, introduction of LTE-M for Internet of Things. But each options has its own constrains, for example femto cells can increase handover signalling, LTE direct has constrains of mobility.

SMNAT based on smart location management and addressing scheme provides an elegant solution to address these basic needs and helps in fostering this new breed of devices. The access network design is simplified with less resource hungry processes associated to mobility management. This is aligned with our endeavour to make the mobile networks cleaner, greener and leaner.

For a typical closed user group type of network, we propose this multiple access mechanism and network topology which will eliminate the involvement of intelligent core network adjuncts in the network area, and to use this intelligent physical layer to directly reach any node over the air interface in the coverage area. Some typical implementations may be automotive communication systems, corporate networks, campus networks, railway networks etc., which use closed numbering scheme and coverage within the community network. This system is 'presence agnostic', hence the network is not aware of the exact location of the mobile node in the network area. Therefore, the mobility management process is not envisaged. For a basic closed user network where the network coverage does not extend beyond 35 km, the design does not follow the cellular topology. The network area is represented by a single Location area.

SMNAT can be proposed for global mobile network and for Device to Device Communication system which is agnostic to the distance between the devices. In such a case, the devices can communicate directly with each other without any physical limitations. The addressing and mobility management will be realised by the physical layer. We would however need non-intelligent wireless access points for boosting the pilot signals throughout the coverage area. We would also require some basic network elements for the purpose of authentication, service authorisation and accounting/billing.

5.4 Conclusion: Destination 2050—The Destiny

It is imperative that with the advancements of telecom technologies, with the growing economic and political might of the social media communication enterprises and the device manufacturers, the connected world of Internet of things, there will be an enormous upwelling of data traffic. Innovative approaches for data offloading need to come up as back haul cost will increase manifold otherwise. Typically for Application to peer (A2P) or P2A applications, local data breakout technologies will facilitate data offload to the nearest break out point towards the internet or cloud. Many of these data offload technologies are available currently. Mobile Edge computing will reduce the processing overhead and complexity in the cloud based mobile networks which will significantly boost the performance, enhance security and reliability especially in IoT and M2M environments. A clear indication that the telecom researchers are moving in that course is the onset of the fog networks, designed to leverage the computing resources of the machine type devices to actuate edge computing. Their aim is to alleviate the core network from some processing tasks and reduce backhaul overhead and latency.

But the problem that still persists is peer to peer communication. At present, the device to device communication technologies like WiFi Direct, LTE direct have constrains on the physical distance between the devices. In case of LTE Direct, which uses the licensed spectrum the maximum distance between the devices could be 500 meters if there are no major physical obstructions between the devices. But these are not complete communication technologies that can actuate network less communication between devices irrespective of the distance between them. If we have to increase the range of communication using LTE direct, then we need to adopt mesh networking, a hop by hop approach where each smartphone contributes to be a hop in between to relay the information from node A to node B. But unless we are able to build up a whole new community of devices to build up this mesh network, these technologies will not be able to bring a fundamental change in the way be communicate today. We still need to depend on 'as is' mobile network for mobility, presence and addressing.

SMNAT is a technology that can drive peer to peer communication regardless of the distance between the devices and location. It can empower the devices to communicate directly with each other at a massive scale eliminating the need of an intelligent network as the routing functions, mobility, addressing, location and presence management can be processed directly by the physical layer. The mobile networks will gradually become redundant in no time.

The impact will be vast. We envisage free and unlimited communication for all. Fuelled by the evolution of such technologies employing smart physical layer addressing and mobility management, the electronics and software required in the device for basic network operations will reduce drastically. This will enable manufacturers to produce even cheaper smartphones and machine type devices. Mobile network will be replaced by clusters of access points to materialise the information superhighway. Mobility will be achieved by smart physical layer mobility management mechanisms. Service discrimination will be actuated by the Over the Top suppliers and no more by the mobile network itself. There may be some mobile operators still left but only providing special services, like high security services for tactical military communications or government officials. Thanks to SMNAT, we have already sowed the seeds of a 'Network-less network'. Having said so, the question that remains is, how much time should we anticipate for the evolution of the network along these lines? Witnessing the incredible development of telecom technologies over the last two decades, we envision that the next 10 years would be adequate to bring about the changes.

So what do we foresee in 2050? With the advancements of biological sciences and neural network technologies, we envisage that by 2050 we would acquire the knowledge to evolve the biological cells to store and process information and execute extrinsic service logic and algorithm. The ability for human beings to communicate within each other without any mobile device, or a wearable should be a reality. These communication capable cells or organs artificially developed could be induced in the newborns which will render the ability to interwork with very low power communication devices implanted permanently in the human bodies. These devices can interact with the nearest access points and actuate communication using the similar physical layer concepts as addressed before. The gamut of thoughts from human mind could be translated to electronic signals and directly communicated over any distance and reciprocated in a similar way via this ecosystem. Telepathy would be the natural way to communicate between 2 human beings separated by distance of any order. It would be no more considered as a paranormal phenomenon. Would you call this the demise of telecom or its rebirth?

Chapter 6
Perspectives on Enabling Technologies for Intelligent Wireless Sensor Networks

Homayoun Nikookar

Abstract Wireless sensor network (WSN) is a network of low-size and low-complexity devices that sense the environment and communicate the gathered data through wireless channels. The sensors sense the environment, and send data to control unit for processing and decisions. The data is forwarded via multiple hops or is connected to other network through a gateway. WSNs have a wide range of applications from monitoring environment and surveillance, to precision agriculture, and from biomedical to structural and infrastructure health monitoring. Technological advances in the past decades have resulted in small, inexpensive and powerful sensors with embedded processing and radio networking capability. Distributed smart sensor devices networked through the radio link and deployed in large numbers provide enormous opportunities. These sensors can be deployed in the air, on the ground, in the vehicles, in or around buildings, bridges, or even on the patient's body.

Wireless communications is the key technology for WSNs. In this chapter important enabling technologies for the development of future wireless sensor networks will be discussed. Technologies such as Distributed beam-forming, Cognitive radio, Joint sensing and communication with OFDM, WSNs Physical Layer security as well as Wavelet technology for context-aware WSNs will be described and their importance in the future developments of WSNs will be discussed.

A typical sensor node in a WSN has a small size (at the size of a button) which has sensors, radio transceivers, a small processor, a memory and a power unit. With the proliferation of wireless sensor networks the requirements on prime resources like battery power and radio spectrum are put under severe pressure. In a wireless environment the system requirements, network capabilities and device capabilities have enormous variations giving rise to significant design challenges. There is therefore an emergent need for developing energy efficient, green technologies that

H. Nikookar (✉)
Netherlands Defence Academy, Den Helder, The Netherlands
e-mail: h.nikookar@mindef.nl

R. Prasad and S. Dixit (eds.), *Wireless World in 2050 and Beyond: A Window into the Future!*, Springer Series in Wireless Technology,
DOI 10.1007/978-3-319-42141-4_6

optimize premium radio resources, such as power and spectrum, even while guaranteeing quality of service. Moreover, many wireless sensor networks operate under dynamic conditions with frequent changes in the propagation environment and diversified requirements. All these trends point to flexible, reconfigurable structures that can adapt to the circumstances and the radio neighborhood. The nodes of future WSNs will most likely be context-aware, heterogeneous with energy harvesting capabilities. In the following sections major enabling technologies for the realization of this perspective on future radio sensor networks are explained.

6.1 Distributed Beamforming for WSNs

A wireless sensor network is formed by radio nodes that are geographically distributed in a certain area, which is shown in Fig. 6.1. The nodes are wireless terminals, or sensors in the network. Here we propose to employ Distributed Beamforming (DB) in the radio sensor network in order to form beams towards the Distant Node. The idea is equally applicable to future Cognitive Radio (CR) network where the nodes of the CR network are able to forward the signals to distant cognitive radio (DCR) node cooperatively. In the context of cognitive wireless sensor network and by adopting the DB method, the CR wireless sensor network increases its coverage range without causing harmful interferences to the already Primary radio services (or Primary Users—PU) operating in the environment and at the same frequency band.

Distributed beamforming (DB) is a cooperative scheme which plays a major role in harnessing the limitation in transmit power of the individual sensor devices of the WSNs. The sensor devices are usually battery driven and are deployed in remote

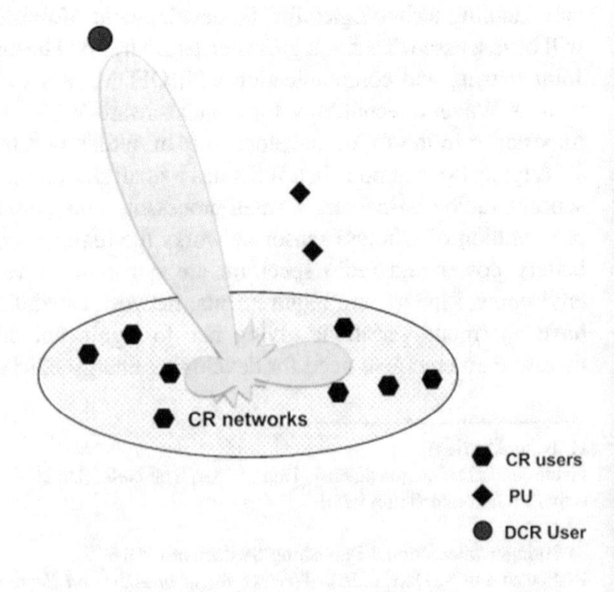

Fig. 6.1 Distributed beamforming for cognitive wireless sensor network [12]

areas where periodic battery replacement is unlikely. Therefore, these devices have to rely on their battery power supply for a longer period of time. In DB, wireless sensor nodes minimize the energy spent per node by cooperatively transmitting the same signal simultaneously so that the signal from each node is added constructively at the receiver. Thus, the gain of the distributed array increases with the number of nodes. This further means that transmit signal power per node decreases by the increasing number of nodes for a fixed bit energy-to-noise power spectral density ratio (E_b/N_0) at the receiver. However, prior to transmission the beamforming nodes have to share information in a coordinated way so that all nodes can transmit the same signal (i.e., synchronization is required). Furthermore, the information exchange during pre-transmission phase creates overhead increase. As a result, the energy consumption of the network increases. Thus, it is critical to choose the right number of nodes to be used in beamforming to optimize the total energy consumption of the network and thus to maintain greenness. The focus of this research should be on optimum node selection and performance evaluation of DB for WSNs [1, 2]. Furthermore, the synchronization requirement and its impact on the system performance should also be investigated. As an application area for this technique the vehicular networks for the Intelligent Transportation Systems (ITS) is suggested as distributed vehicles on the road can collaboratively forward the signal to the distant vehicle with low side lobe interference level. The technique improves communications among vehicles and transport infrastructure which ultimately results in higher vehicles safety and efficient traffic management.

6.2 Cognitive Radio for WSNs

Current WSNs operate in the ISM band. This band is shared by many other wireless technologies giving rise to degradation of performance of WSNs due to interference. WSNs can also interfere other services in this band. The proliferation of WSNs will result in scarcity of spectrum dedicated to wireless sensor communications. Cognitive Radio (CR) technology for WSNs improves sensor nodes communications performance as well as spectral efficiency. It is foreseen that cognitive radio will emerge as an active research area for wireless networks research in the coming years. Unlike conventional radios in which most of components are implemented in hardware, cognitive radio uses software implementations (i.e., Software Defined Radio[1]) for some functionalities enabling flexible radio operation. The radios are reconfigurable and therefore the need to modify existing hardware is reduced. In this context the increasing number of sensing nodes

[1]Software Defined Radio is a software based, programmable and reconfigurable modulation and demodulation technique. With the flexibility that it provides, hybrid platforms can be deployed in the wireless sensor network. By integrating SDR technique in the WSN, with the same (programmable) hardware, more radio standards can be introduced to the network. Therefore, instead of designing again the hardware, only sensor nodes of the WSN are reprogrammed [3].

equipped with wireless communication capability will require faster connectivity and thus, wireless spectrum will have to be adapted to the new requirements (of bandwidth). Cognitive radio will prevent the need to implement hardware upgrades with emergence of new protocols in the future. It will allow the cognitive radio-enabled nodes to search for the best frequency-based pre-determined parameters. It should be noted that CR-WSNs differ from conventional WSNs in several aspects. One of important issues in this regard is interference to other wireless networks or Primary Users (PUs) [4, 5]. Protecting the right of Primary users is the major concern of the CR-WSN. Therefore, miss detection probability of PUs should be minimized in order to minimize interference with the PUs. False alarm probability should also be minimized as large false alarm rates cause spectrum to be under-utilized. High false alarm and miss detection probability in CR-WSNs result in a long waiting delay, frequent channel switching and significant degradation in throughput. In the context of future WSNs research these issues have to be investigated in detail.

Applying CR technique to mobile or dynamic WSNs is challenging. One major example of mobile WSNs is the vehicular networks. Adaptive and reconfigurable Vehicle-to-Vehicle (V2V) and Vehicle-to-Infrastructure (V2I) have numerous applications (such as emergency warning systems of vehicles, Cooperative adaptive cruise control, Cooperative forward collision warning, intersection collision avoidance, Highway-rail intersection warning, etc.). The flexibility and the agility offered by cognitive radio is very useful for resilient communication among mobile wireless nodes (cars). For example in platooning of automated vehicles of future or in the emergency situations where a cognitive radio network should be reconfigured in real time by trading bandwidth maximization or power minimization for more resiliency. Major challenges in this regard are Quality of Service guarantee (e.g., low delay, high reliability in high mobility scenarios) and its trade off with bandwidth. Furthermore, the unique features of the mobile WSNs such as mobility of nodes, topology change and cooperation opportunities among nodes need to be taken into account. Other challenging topic in this direction is dynamic spectrum access for mobile WSNs. In this regard and specifically the spectral holes for mobile CR-WSNs working in highway scenarios are important as highways (unlike downtown and urban areas) are open spaces and there is a high chance of finding a spectrum hole that can be used opportunistically. This in turn can answer some of bandwidth and congestion problems of the network.

In wireless sensor networks typically the nodes have low-height antennas. In these applications the radio propagation channel characteristics and among others the path loss exponent is considerably different from the free-space channel. Therefore, routes with more hops and with shorter hop distances can be more power-efficient than those with fewer hops but longer hop distances. To this end carrying out research on CR-WSN channel and investigating the impact of dynamic channel on the adaptive self-configuration topology mechanism of sensor network is suggested to gauge to which extent this can reduce energy consumption and increase network performance.

The research on CR for WSN positions well in the scope of future developments of WSNs as cognitive radio will be the enabling technology for wireless sensor networks that have to coexist with other networks [6] and have heterogeneous devices. The reconfigurability, adaptation and interoperability capabilities of the cognitive radio are major accents of this technology which are pronounced in the research perspectives of future WSNs.

6.3 Joint Communication and Sensing in One Technology for WSNs

As is clearly mentioned in the first Conasense book [7] a very important integration strategy concerns communications, positioning and sensing systems. This integrated vision involves an "active" integration with new business opportunities able to merge three worlds—communications, navigation and local/remote sensing—that have been apart for years. This vision is the focus of the Communications, Navigation, Sensing and Services (CNSS) paradigm. In CNSS, communications, navigation and sensing systems can mutually assist each other by exploiting a bidirectional interaction among them (see Fig. 6.2) [8].

Integrated sensing and radio communication systems have emerged in sake of system miniaturization and transceiver unification. With the current technological

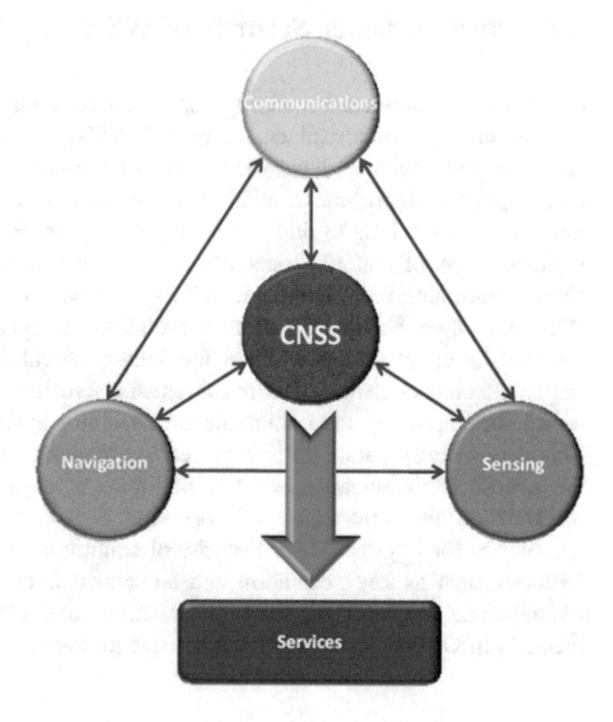

Fig. 6.2 Integration of communications, navigation, sensing and services [8]

advancements the radio frequency front-end architectures in sensing networks and radio communications become more similar. Orthogonal Frequency Division Multiplexing (OFDM) as a capable technology already and successfully used in wireless communications (e.g., in IEEE802.11a,g,n,p) can be used in wireless sensor networks. Among the state-of-the art transmission schemes the seminal concept of [9] on joint ranging (location) and communication using OFDM technology is important. In this context OFDM can find applications in future WSNs where context-awareness (location information) is important and, in addition to sensing, data exchange among nodes or between nodes and fusion center is essential. It is worth mentioning at this juncture that:

i. In addition to reconfigurability and adaptation capability of OFDM, by using this technology besides achieving both functionalities simultaneously, the bandwidth will be efficiently used.
ii. By using OFDM the Doppler effect can be detected in real time, which may be used to correct the highly mobile communication channels for high-speed data link of V2I or V2V of Vehicular Sensing Networks. Another application is a network of UAVs in the air which senses the environment. In this scenario the UAV nodes are highly mobile and due to the high speed regime the Doppler should be estimated and compensated. OFDM is a capable technology in this regard.

6.4 Physical Layer Security of WSNs

Security of communication among nodes and between nodes and fusion center is one of the most important challenges of WSNs. Generally speaking, computer scientists and engineers have tried hard to continuously come up with improved crypto-graphic algorithms to address the security issue challenge. But typically it does not take too long to find an efficient way to crack these algorithms. With the rapid progress of computational devices, the current cryptographic protocols are already becoming more unreliable [10]. A new paradigm which is proposed is the 'Physical Layer Security'. Unlike the traditional cryptographic approach which ignores the effect of the wireless medium, physical layer security exploits the important characteristics of wireless channel (such as fading, interference, and noise) for improving the communication security against eavesdropping attacks. This new security paradigm is expected to complement and significantly increase the overall communication security of future wireless sensor networks. Further research on the Information Theory and Signal Processing techniques, and approaches for Physical Layer Security of communications in WSNs is envisaged. Methods such as key generation and agreement over wireless channels, various multiantenna (MIMO) transmission schemes, and efficient resource allocation methods in OFDM systems are a few ideas to name.

6.5 Wavelet Technology for Context Aware and Reconfigurable WSNs

As already mentioned in WSNs several nodes operate at the same frequency band in a network. They share the spectrum, and therefore may interfere each other. Accordingly, in these networks the interference becomes a major problem to be addressed. In this research direction the context aware design of communication signals for adaptive wireless sensing networks is suggested to cope with the interference in an intelligent way. On the other hand the Wavelet packet transform has recently emerged as a novel signal design technique with attractions such as good time–frequency resolution, low sidelobes, and the reconfigurability capability [11]. The wavelet approach is advantageous for the signal design of smart wireless sensor networks [6] mainly because of its flexibility, lower sensitivity to distortion and interference, and better utilization of power and spectrum. For the minimum interference to the adjacent bands, the wavelets should be maximally frequency selective. Commonly known wavelets are not frequency selective in nature and hence result in poor spectrum sharing performance. To alleviate this problem, and in the scope of interference problem of WSNs the design of a family of wavelets that are maximally frequency selective in nature is suggested. To this end, the design constraints should be first enlisted. Then the challenging problem which is most likely non-convex, should be first reformulated into a convex optimization problem and should be solved using Programming tools. Before implementation of this technology first through simulation studies the benefits of such a newly designed wavelets for a context-aware wireless sensing networks should be demonstrated.

It has to be mentioned that, in addition to interference, the wavelet design framework of this research will be easily applied to other design criteria (e.g., low power and greenness of the network, throughput or latency performance, or timing error or synchronization constraint of WSNs) by merely altering the objective function of the design procedure. However, to be able to do so, the desirable properties of the wavelet bases must be translated into realizable objective functions. This can at times be challenging.

Wireless sensor networks work on the principle of adaptive distributed load sharing amongst the constituent nodes. They also exploit spatial correlation between the data collected between nodes that are physically close together. Wavelets can also be used to exploit spatial correlation between the data collected between sensing nodes that are physically close together. Accordingly, distributed multiresolution algorithms are suggested to be developed to reconstruct the information gathered by the nodes with the sensors spending as little energy as possible. In this research direction wavelets are designed for adaptive distributed processing algorithms in large wireless sensor networks for power efficient data gathering through use of spatially correlated data. A wireless network architecture is suggested which will efficiently support multi-scale communication and collaboration among sensors for energy and bandwidth efficient communication, while reducing communication overhead, and saving energy.

Furthermore, with regard to the capability of wavelets in signal compression and particularly its importance for the big data challenge of WSNs and in mitigating the burden of the storage utilization as well as in mitigating data congestion, research and development on data compression capability of wavelets used for the future WSNs in which data is compressed before it is sent out, is suggested. Moreover, due to nature of wavelets, the technique will be beneficial not only in reducing White noise (i.e., de-noising) but also in mitigation of wide range of other interferences such as partial discharge, corona, lightning and interference in smart energy networks, in particular, or emission of other signals and interferences existing in WSNs working in industrial environments, in general.

6.6 Intelligent Wireless Sensing in 2050 and Beyond

With regard to the proliferation of wireless sensing networks in the coming years and decades the big data issue is expected to emerge as a remarkable challenge for the WSNs of the future. Mitigating the burden of the storage utilization as well as mitigating data congestion, research and development on data compression techniques and particularly wavelets, for the future WSNs (in which data is compressed before it is sent out), is suggested. Noticing the rapid deployment of Internet-of-Things (IoT) and WSNs in smart cities, homes, factories, hospitals, cars, and wide spread applications of wireless sensing technologies in medical, industrial structural health monitoring, climate change monitoring, earthquake sensing, agricultural and food chain monitoring and control, etc., and seeing the imminent big data challenge on one side and the excellent data compression capability of wavelet on the other side, it is indeed expected that the wavelet technology will emerge as *the* key technology in the big data hype of WSNs in the coming period.

Enabling technologies for wireless sensor networks, among others those discussed in this chapter, will pave the way for full realization of cooperative, adaptive, reconfigurable, context aware, cognitive, secure and green wireless sensing networks of the future.

6.7 Conclusion

In this chapter the remarkable enabling technologies for the realization of future intelligent WSNs were discussed. Flexible software structures will be needed to reconfigure the signal setting and implement the adaptation of the designed signal for the intelligent WSNs. Advanced technologies to face major challenges of future

WSNs such as coexistence with other networks and interference, reconfigurability, low power, and high spectral efficiency were addressed. These technologies constitute the future strategic research agenda on WSNs.

References

1. Nikookar H (2013) Green wireless sensor networks with distributed beamforming and optimal number of sensor nodes. In: Ligthart, LP, Prasad R (eds) Communication navigation sensing and services. River Publishers. ISBN: 13-9788792982391
2. Lian X, Nikookar H, Ligthart LP (2012) Distributed beamforming with phase-only control for green cognitive radio networks. EURASIP Journal on Wireless Communications and Networking 2012, 1687-1499
3. Li Y (2009) Integrating software defined radio into wireless sensor network. Thesis, Royal Institute of Technology (KTH)
4. Budiarjo I, Nikookar H, Ligthart LP (2010) Modulation techniques for cognitive radio in overlay spectrum sharing environment. In: McGuire PD, Estrada HM (eds) Cognitive radio: terminology, technology and techniques. Nova Science Publishers, NY, USA, pp 73–99. ISBN: 978-1-60876-604-8
5. Budiarjo I, Nikookar H, Ligthart LP (2008) Cognitive radio modulation techniques. IEEE Signal Process Mag 25:24–34
6. Nikookar H (2015) Signal design for context aware distributed radar sensing networks based on wavelets. IEEE J Sel Top Sign Process 9(2):204
7. Ligthart LP, Prasad R (eds) (2013) Communications, navigation, sensing and services (CONASENSE) book. River Publishers, Denmark. ISBN 9788792982391
8. http://www.conasense.org
9. Genderen PV, Nikookar H (2006) Radar network communication. In: 6th international conference on communications, Bucharest, Romania, pp 313–316
10. Zhou X, Song L, Zhang Y (eds) (2014) Physical layer security in wireless communications. CRC Press
11. Nikookar H (2013) Wavelet radio: adaptive and reconfigurable wireless systems based on wavelets. Cambridge University Press
12. Lian X (2013) Adaptive and distributed beamforming for cognitive radio. PhD thesis, Delft University of Technology

Chapter 7
Energy Efficient Joint Antenna Techniques for Future Mobile Communication: Beyond 2050

Bo Han

Abstract The future wireless communication will be a cloud network with everything be connected; as the trend of internet of everything, the cellar becomes smaller and servers even for very tinny terminals. Such a system must be both spectrum and energy efficiency due to the limited resources; also the size of the network elements must be smaller to fit the dimension requirement. To deal with those challenges, this chapter presents a joint antenna technique that can be used for the next generation mobile communication: by exploring higher order beam space MIMO with joint antenna techniques, it seems to some extent, it's possible to reach similar performance as conventional MIMO while using half of the energy and smaller device dimension.

Keywords Beam space MIMO · Single RF · ESPAR · Energy efficiency · 16-QAM · Pattern mapping

7.1 Introduction–The Ubiquitous Connectivity to Humans

Since the first telegraph has been transmitted, the mobile communication started its development aiming at bringing human together and making the society smaller. With the help of advances in electronic circuits and micro electronic chipsets, the mobile communication has made huge success in the past dozen of years. New standard has been proposed and large amount of efforts have been invested by academic and industry. As a result the mobile network has been evolved from GSM to UMTS, HSPA, LTE, LTE-A, and keep on moving.

However, the way of the interaction between human and the society has been expanded by all kinds of methods in the recent years: the mobile telephone itself is

---

B. Han (✉)
HUAWEI Technologies, Amsterdam, The Netherlands
e-mail: bohan@huawei.com

R. Prasad and S. Dixit (eds.), *Wireless World in 2050 and Beyond: A Window into the Future!*, Springer Series in Wireless Technology,
DOI 10.1007/978-3-319-42141-4_7

no longer a simple device just to make voice calls, instead, it is a device that can make the personal living information retrieved at anytime and anywhere. The habit of users that addicts on the connected world make a trend, that is the higher capacity higher data rate demand on the mobile communication within a compact size.

Also, Spectrum is a nonreproductive resources that must be carefully used, thus higher order modulation with MIMO is inevitable; meanwhile, the future ubiquitous connectivity requires a always online model of the devices, thus the energy efficiency of those battery powered devices must be carefully designed, as the terminals also have limitation on the dimension. The base stations nowdays has been evolved from the macro station to micro station, and lampsite as well as pico station, in future it will be even smaller but with more advanced antenna techniques: as a result, the multi standard carrier aggregation, multi-band MIMO and massive MIMO on a small pico station will be quite challengeable for antenna design.

7.2 Low Complexity MIMO Communications

To satisfy the demand for high data rate, within a crowded bandwidth, the MIMO technique was proposed to improve spectral efficiency and the radio link performance. When MIMO communication is used, the radio link enjoys the benefits of spatial multiplexing and diversity gain, such as better spectrum efficiency, lower error probability, larger system capacity, etc. [1, 2].

However, the conventional MIMO approach has some inherent disadvantages, such as, it requires more than one radio frequency chains for each antennas. The radio frequency chain contains power amplifier, mixer, frequency synthesizer, analog to digital converter, etc., which burden the circuit complexity. Moreover, these circuit components require additional power consumption. When large numbers of antennas are used in MIMO techniques to support large capacity and high data rate [3, 4], the circuit power consumption can reach very high levels, which hinders the application of MIMO techniques on mobile devices.

The compact size is another constrain for MIMO techniques to be used on mobile devices. In order to have independent fading for each MIMO data stream, the correlation among antenna elements should be as small as possible. This leads to larger antenna spacing requirement, e.g. 2λ among the elements, which is very difficult to be accommodated on compact sized devices.

Thus the concept of low cost, low complexity communication was proposed, which aims at building a communication system with higher data rate and better spectrum and energy efficiency, but also with low complexity and compact size of mobile devices in the system. The conventional MIMO techniques give good promise in terms of high data rate and good spectral efficiency, but the compact size and energy efficiency was still a challenge for mobile (portable) devices.

7.2.1 The Beam Space MIMO

Actives of the academic and industries in this area: with the aforementioned requirements and constrains, an improved MIMO transmission technique was proposed in [5], where the MIMO data streams were mapped to different orthogonal radiation patterns using a compact antenna, instead of letting them go through several different active antennas. Those orthogonal basis patterns make the fading independent on each MIMO data stream, so that these data streams can be considered as if they are from a conventional MIMO transmitter with multiple antennas.

This mechanism is the so-called beam space MIMO transmission, where the beam pattern synthesization was done by using the Electronic Steerable Parasitic Array Radiators (ESPAR) [6], which is composed of one active element with several parasitic elements spaced by λ/N (λ is the wavelength, N is an integer). Moreover, only one active element is required in such beam space MIMO transmission; the radiation patterns are altered by changing the load values on the parasitic elements.

Figure 7.1 gives the basic concept of the beam space MIMO transmission, where two MIMO data streams w_0, w_1 are mapped onto two basis patterns Φ_0, Φ_1, and will be transmitted independently through the channel [7].

The previous research towards this direction has proved that such beam space MIMO can only support up to PSK modulations [8] and had only considered pure imaginary loads [9, 10].

To waive these constrains and extend the capability of ESPAR antenna for beam space MIMO transmission with more complicated constellations, such as 16-QAM; also with *complex* adjustable loads are considered and proposed in this section. Further more, a two-element ESPAR architecture is proposed to further reduce the antenna size.

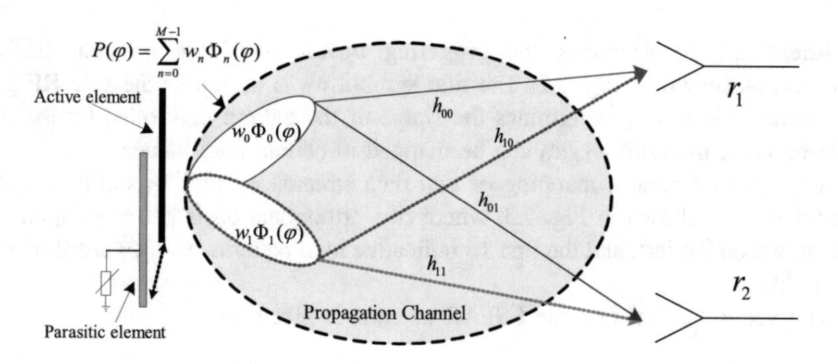

Fig. 7.1 Beam space MIMO symbols mapping

7.2.2 Beam Space MIMO with 16-QAM

The radiation pattern of such a beam space MIMO system is a composition of the weighting coefficient and the basis patterns. for the 2D pattern, the total pattern of the beam space MIMO over an ESPAR antenna with M elements is expressed as a linear combination of M basis patterns and MIMO data stream w_n as:

$$P(\varphi) - \sum_{n=0}^{M-1} w_n \Phi_n(\varphi) \tag{7.1}$$

For an ESPAR antenna with 2 elements, the basis patterns are calculated through a Gram-Schmidt orthogonal procedure as described in [11]:

$$\begin{aligned} \Phi_0(\varphi) &= 1/k_0 \\ \Phi_1(\varphi) &= \left(e^{jb\cos\varphi} - 2\pi I_0(jb)/k_0^2\right)/k_1 \end{aligned} \tag{7.2}$$

where the Gram-Schmidt constants k_0, k_1 and the weighting coefficients w_n (which are also MIMO symbols to be mapped) in (7.1) are given by:

$$\begin{aligned} w_0 &= i_0 k_0 + 2\pi i_1 I_0(jb)/k_0 \\ w_1 &= i_1 k_1 \end{aligned} \tag{7.3}$$

where i_0, i_1 is the current on the corresponding antenna element, $b = 2\pi d, d$ is the inter-element distance normalized to wavelength and $I_0(x)$ is the zero-th order modified Bessel function of the first kind.

For the beam space MIMO transmission, recalling (7.1), the pattern becomes:

$$P(\varphi) = w_0\left(\Phi_0(\varphi) + \frac{w_1}{w_0}\Phi_1(\varphi)\right) \tag{7.4}$$

Equation (7.5) describes the triggering operation of the transmit ESPAR antenna, as shown in Fig. 7.2. The first symbol w_0 is driven to the sole RF port, while the ratio w_1/w_0 determines the shape of the pattern controlled by the load values. Thus, the ratio w_1/w_0 can be mapped to certain load values.

The azimuth pattern mapping of two data streams w_0 and w_1 (with 16-QAM modulation) is shown in Fig. 7.3, where two orthogonal basis patterns Φ_0 and Φ_1 are shown on the left, and the first 16 indicative total patterns of $P(\varphi)$ are shown on the right.

The vector current $\mathbf{i}$ on the ESPAR antenna is given by:

$$\mathbf{i} = (\mathbf{Z} + \mathbf{X})^{-1}\mathbf{v} \tag{7.5}$$

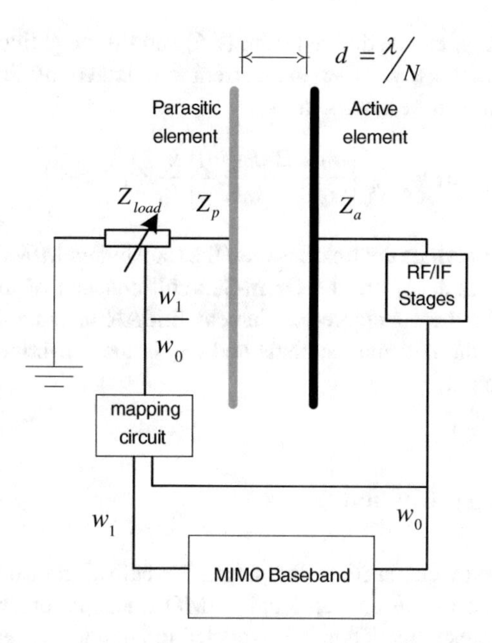

Fig. 7.2 Single RF MIMO transmission

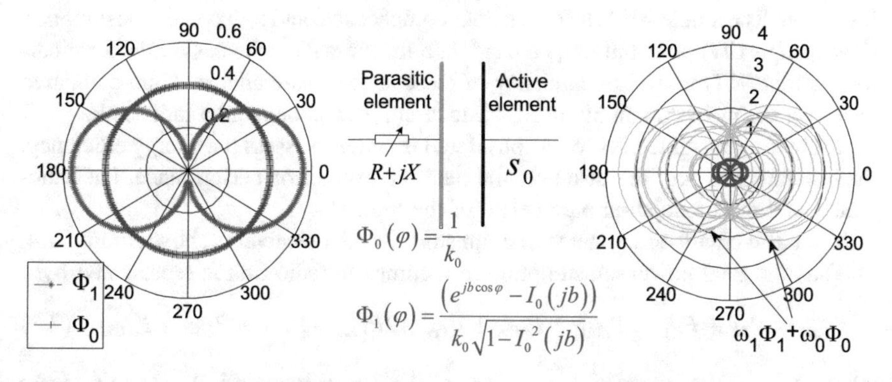

Fig. 7.3 16-QAM MIMO symbol mapping to basis patterns for 2 elements ESPAR

where $\mathbf{Z}$ is the antenna impedance matrix, $\mathbf{v} = [\,u_s \quad 0\,]^T$, and the $\mathbf{X} = diag[\,50 \quad x_1\,]$ is the diagonal load matrix. Thus, the current distribution $\mathbf{i} = [i_0, i_1]^T$ and the parasitic load x_1 can be expressed in the following equation:

$$\left(\begin{bmatrix} Z_{11} & Z_{12} \\ Z_{21} & Z_{22} \end{bmatrix} + \begin{bmatrix} 50 & 0 \\ 0 & x_1 \end{bmatrix}\right)[\,i_0 \quad i_1\,]^T = [\,u_s \quad 0\,]^T \tag{7.6}$$

where Z_{11}, Z_{22} are the self-impedance of the active element and parasitic element respectively; the mutual coupling impedances are identical ($Z_{21} = Z_{12}$), so the parasitic load value x_1 can be derived from (7.6) and the weighting coefficients (also the MIMO symbols) w_0, w_1 and the current are linked by Eq. (7.3), thus the required load value can be rewritten as:

$$x_1 = -\left(\left[\frac{w_0}{w_1} - \frac{2\pi I_0(jb)}{k_0 k_1} \right] \frac{k_1}{k_0} Z_{21} + Z_{22} \right) \quad (7.7)$$

Equation (7.7) provides the link between the transmitted MIMO symbols and the required load values; k_0, k_1 are the Gram-Schmidt constant of the two orthogonal patterns Φ_0, Φ_1, for the considered 2-element ESPAR antenna. Z_{21} is the mutual impedance among the antenna elements and Z_{22} is the self-impedance of the parasitic antenna element.

7.3 The Energy Efficiency

As shown in the previous section, the single RF MIMO transmission has similar system performance as the conventional MIMO transmission, with only one RF chain used at the transmitter. Thus it is expected to be more power efficient than the conventional MIMO system. As the trend of ubiquitous connectivity, where a lot of terminals are small size and powered by batteries [12]. Thus the energy efficiency is important. The single RF MIMO has both compact size and low power consumption (low complexity) potential to be a candidate for the radio schemes on the terminals for Internet of Things. The efficiency of these schemes are analyzed and compared based on the most commonly used, basic circuit components in a radio link.

There are several factors to be considered in order to assess the energy efficiency. Such parameters are: the distance, the data rate, the error performance, the bandwidth, and the modulation parameters of the link, etc.

The basic circuit components of a transmitter and a receiver is shown in Fig. 7.4. Thus the total power consumption in a complete radio link is represented by:

$$P_{total} = P_{TX} + P_{RX} \simeq P_{DAC} + P_{MIX} + P_{PA} + 2P_{PLL} + P_{MIX} + P_{LNA} + P_{ADC} \quad (7.8)$$

where P_{TX} is the power consumption of the transmitter and P_{RX} is the power consumption of the receiver. In order to simplify the model, the base band power consumption is considered to be very small compared to the main circuit component in a radio link.

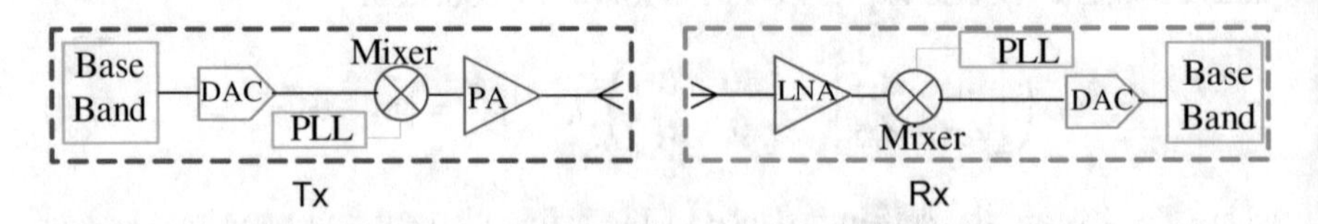

Fig. 7.4 Basic circuit components in a transmitter (*left*) and a receiver (*right*)

In order to keep a stable radio link, (i.e., for the error performance to be kept at a certain level), the power amplifier consumes dynamically due to linearity and modulation depth. Thus rename of the power consumption of the amplifier as P_T, which denotes a transmission-related power consumption, where the $P_T = P_{PA}$ in Eq. (7.8). The rest of the circuit component part is assumed with power consumption P_C.

The energy efficiency is defined as the minimal energy required in order to transfer one bit, assuming the information data length is L, the data rate of the radio link is R_b, Thus the duration of the transmission is given by $T_{on} = L/R_b$, and the energy efficiency is given by [13]:

$$E_{bt} = (P_T + P_C)/R_b \tag{7.9}$$

The transmission power P_T includes the transmission power to the air $P_{T(air)}$ and the self power consumption $P_{A(self)}$ of the power amplifier. It is noted as $P_T = P_{T(air)} + P_{A(self)} = (1+\alpha)P_{T(air)}$, where α is given by $\alpha = \xi/\eta - 1$, and ξ, η are the linearity factor and power drain efficiency of the power amplifier, respectively. For M-ary signaling, ξ is related to the Peak to Average Power Ratio (PAPR) given by $\xi = 3(M - 2\sqrt{M} + 1)/(M-1)$ [13].

The transmission power to the air $P_{T(air)}$ can be calculated through the radio link budget. According to [13], the energy budget is shown as:

$$P_{T_{Air}} = \overline{E_b} \times R_b \times \left(\frac{4\pi}{\lambda}\right)^2 \times \frac{U_{link} N_F}{G_t G_r} \times d^{\kappa} \tag{7.10}$$

where $\bar{E}_b$ is the required energy per bit for a given error probability $\bar{P}_b$, and R_b is the data rate noted by bit/s. Energy efficiency of the circuit part is reduced by the data rate. Also, a higher constellation requires more power to overcome the error performance and compensate the linearity of the power amplifier.

The theoretical error probability for BPSK signaling is given by:

$$\overline{P}_b = Q(\sqrt{2\gamma_b}) \tag{7.11}$$

where γ_b is given by $\gamma_b = \overline{E}_b/N_o$ in a SISO link; for MIMO link with the Alamouti scheme, $\gamma_b = (|\mathrm{H}|_F^2 \cdot \overline{E}_b)/(M_t \cdot N_o)$, where $|\mathrm{H}|_F^2$ is the squared Frobenius norm of the MIMO channel matrix H, and it is assumed that the transmission power is assumed to be equally allocated on the M_t antennas. According to the Chernoff bound and approximating this bound as an equality [13], thus the required energy per bit $\overline{E}_b$ (J/bit) can be approximated as:

$$\overline{E}_b = \frac{M_t \cdot N_o}{\overline{P}_b^{\frac{1}{M_t}}} \tag{7.12}$$

where M_t is the number of transmission antennas. For higher order signaling, e.g., M-ary signaling (i.e., 16-QAM), the error probability is worse than BPSK [14], following the same procedure as in (7.12), the required energy per bit for M-ary signaling with constellation size b is given by [13]:

$$\overline{E}_b = \frac{2}{3}\left(\frac{4(1-1/2^{b/2})}{b\bar{P}_b}\right)^{\frac{1}{M_t}} \frac{M_t \cdot N_o(2b-1)}{b} \tag{7.13}$$

where the b is the constellation size, and R_b is the data rate for binary signal, which is the same as the bandwidth $R_b = B$. Figures 7.5 and 7.6 give the energy efficiency of the radio link considering both transmission power consumption and circuit power consumption. For SISO link $M_t = 1$; when more than one receiving antenna is used, the circuit energy P_c increases while the required transmission energy to the air reduces by approximately N_r. For single-RF transmission, only one RF chain is used, thus the transmission energy is further reduced and the circuit energy on the transmitter is also reduced.

With the aforementioned assumptions, the energy efficiency is plotted in Figs. 7.5 and 7.6, where the comparison of the SISO link with different constellation sizes is provided in Fig. 7.5; the energy efficiency of a BPSK MIMO link is better than a SISO link at larger distance. The energy efficiency of the MIMO link is shown in Fig. 7.6, where it can be concluded that with the help of MIMO link, the higher order modulation(16-QAM) can have better energy efficiency at a larger distance compared with that of a SISO link.

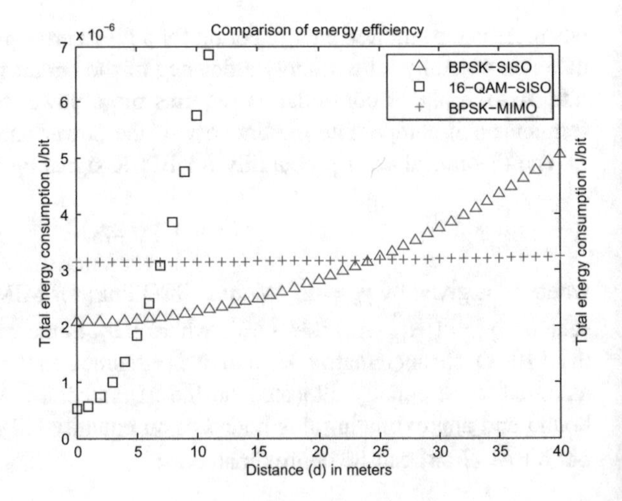

Fig. 7.5 SISO link with $R_b = 100$ Kb/s

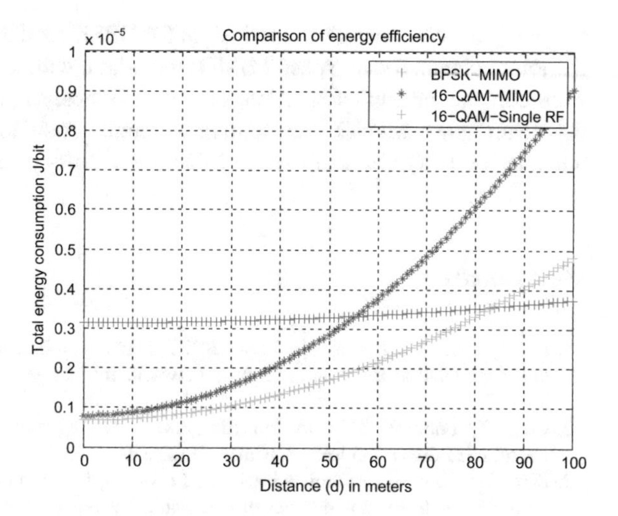

Fig. 7.6 MIMO link with $R_b = 100$ Kb/s

7.4 Balance Between Energy Efficiency and Security

Network privacy and security is a key issue for the network in 2050. with the help of soft defined radio, the network security can be implemented on almost on physical layer, while the network coding and network layer encryption and application layer authentication should also be well designed. changing the network layer protocol can some how influence on the security.

One of the directions for energy efficiency is to simplify the protocols, especially for simple devices like sensor nodes. When the unnecessary re-transmission mechanism is reduced, the percentage of useful information bit over total message bit increases, which in turn increases the energy efficiency and extend the battery life.

However, such kind of simplifying the protocols leads to an bottleneck: the security issue. When the ubiquitous connectivity links everything to the cloud, any breakpoint in the net has the possibility that leads to a potential danger for the whole network.

Here this section mainly analyzes the energy efficiency on the physical layer, while keeps the network layer structure unchanged.

7.5 Conclusion

As the mobile communication going to ubiquitous connectivity, a larger data rate and higher throughput is necessary, also the base station becomes smaller, the antenna dimension has to be small for the advanced MIMO techniques. inspired by the previous work that gives the insight for BPSK modulation, this chapter presents the joint antenna (beam space MIMO) technique with higher order modulation and even compact antenna size. Also, the quantify analysis of the energy efficiency on the beam space MIMO with among different modulations verifies that the joint antenna for 16-QAM beam space MIMO is a energy efficient solution.

References

1. Han B, Kalis A, Papadias CB, Prasad R (2013) Energy efficient MIMO transmission with high order modulation for wireless sensor network. In: European signal processing conference, pp 1–5
2. Chan T, Popovski P, Carvalho ED, Sun F (2013) Diversity-multiplexing trade-off for coordinated direct and relay schemes. IEEE Trans Wireless Commun 12(7):3289–3299
3. Kim TM, Sun F, Paulraj AJ (2013) Low-complexity mmse precoding for coordinated multipoint with per-antenna power constraint. IEEE Signal Process Lett 20(4):395–398
4. Sun F, Carvalho ED (2012) A leakage-based mmse beamforming design for a mimo interference channel. IEEE Signal Process Lett 19(6):368–371
5. Kalis A, Kanatas AG, Papadias CB (2007) An ESPAR antenna for beamspace MIMO system using PSK modulation schemes. In: IEEE international conference on communications, pp 5348–5353
6. Kalis A, Kanatas AG, Papadias CB (2008) A novel approach to MIMO transmission using single RF front end. IEEE J Sel Areas Commun 26:972–980
7. Alrabadi ON, Papadias CB, Kalis A, Prasad R (2009) A universal encoding scheme for MIMO transmission using a single active element for PSK modulation schemes. IEEE Trans Wireless Commun 8:5133–5142
8. Alrabadi ON, Kalis A, Papadias CB, Kanatas AG (2008) Spatial multiplexing by decomposing the far-field of a compact ESPAR antenna. IEEE PIMRC, pp 1–4
9. Alrabadi ON, Perruisseau-Carrier J, Kalis A (2012) MIMO transmission using a single RF source: theory and antenna design. IEEE Trans Antennas Propag 60:654–664
10. Alrabadi ON, Divarathne C, Tragas P, Kalis A, Marchetti N, Papadias CB, Prasad R (2011) Spatial multiplexing with a single radio: proof -of- concept experiments in and Indoor environment with a 2.6GHz prototype. IEEE Commun Lett 15:178–180
11. Barousis VI, Kanatas AG, Kalis A (2011) Single RF MIMO systems: exploiting the capabilities of parasitic antennas. In: IEEE 74th Vechicular Technology Conference, pp 5–8
12. Bhardwaj M, Chandrakasan AP (2010) Bounding the life time of sensor networks via optimal role assignments. In: Proceedings of INFOCOM, vol 3, pp 1587–1596
13. Cui S, Goldsmith AJ, Bahai A (2004) Energy efficiency of MIMO and cooperative MIMO techniques in sensor networks. IEEE J Sel Areas Commun 22:1089–1098
14. Cui S, Goldsmith AJ, Bahai A (2003) Modulation optimization under energy constraints. In: IEEE International Conference on Communications, ICC ’03, vol 4, pp 2805–2811

Chapter 8
The Software Defined Car: Convergence of Automotive and Internet of Things

Mahbubul Alam

Abstract By the year 2020 an entire new generation would have grown up with mobile phone, high speed broadband, cloud services and apps for everything on smart device and at their fingertips, they are the "digital natives". This is mainly due to the megatrend "digitalization" of services fueled by the Internet of Things (IoT), which has the power to disrupt every industry and everything around us. It will completely change the way we live, play, work, pay, travel and drive. IoT is rapidly transforming the automotive industry from connected car to full autonomous car, enabling new consumption and monetization models such as car sharing, pay-as-you-drive, etc. 5G systems have a key role to play in order to make the vision of the full autonomous car a reality however there are a few challenges it will have to address in order to accommodate all the autonomous driving use cases discussed in this chapter.

Smartphone are not only a need in today's society but essential to our daily lives. The emergence and popularity of IoT are due to the fact that the upwardly mobile urban populace across the world feels a strong lifestyle need for connected devices. The newer technologies, today, will enable users to stay connected through a range of devices, all through IoT. We are increasingly witnessing an IoT-driven change across industries. Industries such as those in transportation, retail, oil and gas, manufacturing and healthcare will see enormous benefits. For instance, supply chains will soon have provision to track automobile parts and materials in real-time. This is bound to reduce work capital requirements and avoid disruptions in manufacturing. Likewise, the healthcare sector will be able to remotely monitor its staff and patients. The oil and gas industry, through smart components, will see reduced operations cost and higher fuel efficiency. Companies in manufacturing industry

M. Alam (✉)
Movimento Group, Sunnyvale, California, USA
e-mail: Mahbubul.Alam@movimentogroup.com

R. Prasad and S. Dixit (eds.), *Wireless World in 2050 and Beyond: A Window into the Future!*, Springer Series in Wireless Technology,
DOI 10.1007/978-3-319-42141-4_8

will see speedy responses to demands, using smart sensors and various digital control systems. Cities and society will also see a lot of societal and environmental benefits, where pollution and traffic will be reduced considerably. As we shift our focus to the automotive industry we cannot fail to notice the way IoT is engulfing the space. However, the automotive industry remains low-profile and calls for rapid transformation as compared to other industries.

Also, the automotive industry aims to eliminate possible and probable fatalities and accidents using technologies that make autonomous driving a reality. Today, cars are no more mere machines that drive users from point A to point B. Cars are expected to be as functional as homes and workplaces if not more. Users are demanding easy access to data and services that they have on their phone, in their cars.

The author of this chapter predicts "The vehicle of tomorrow will be a super-computer on wheels internally connected through a deterministic IP network where the software features will determine the car's behavior on the roads and the vehicle will work for you, ushering in an era of the Software Defined Car".

This presents new business possibilities for content providers, mobile platform providers and OEMs. With new horizons opening up for the industry at large, strategic decisions of today will have an impact tomorrow. In other words, we are paving the future business pathway by the strategic choices we make today, in the automotive industries. The returns may come in small packages, but the moves have to be keenly strategic.

At the end, from the user's point of view, it is all about delivering value as we enhance the in-car experience. Heartening news is that the auto industry is already rapidly transforming, offering connected services and the driving force is the amalgamation of technologies under the umbrella of IoT.

8.1 Convergence of Automotive with Internet of Things

The automotive industry is fast introducing innovative features on the basis of consumer trends, preferences, safety bulletins and market positioning. Design engineers are burning the midnight oil to make IoT a potent enabler in the new era of automotive business. Processors and networking devices are set to take the industry a notch higher, provided the technological minds leverage it strategically and to the hilt. Tesla CEO, Elon Musk said that in 20 years, owning a car will be a lot like owning a horse [1] and Morgan Stanley expects "full automation" by 2022, creating $1.3 trillion value in the United States alone.

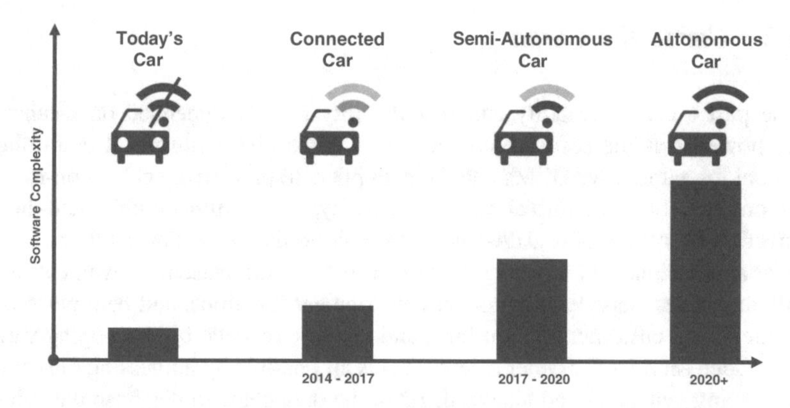

Fig. 8.1 Software complexity for automotive industry

8.1.1 Within the Car

Tomorrow's opportunities lie in what we call the 'Software-Defined Car', a connected vehicle replete with sensors and processors that rely on software which are running 30–100 microprocessors known as Electronic Control Units (ECUs) interconnected with each other through in-car networks. In 2016, a typical high-end car like an S-class Mercedes Benz runs over 100 million lines of software code—compare that with an F-22 running about 1.7 million, and Boeing's new 787 Dreamliner running about 6.5 million lines of code. As the cars of today evolve towards full autonomous functionalities, it is expected that software in the car will grow rapidly to about 500–700 million lines of code as seen from Fig. 8.1 and the complexity of managing software will grow exponentially [2]. Rise of wireless connectivity and software complexity in cars will require OEMs, and their suppliers to integrate a software management platform to address bug fixes, security patches, software/firmware updates, and to introduce new feature/functionality like Tesla's Autopilot using car wireless connectivity, commonly referred as Over-The-Air (OTA). Car OTA must be performed securely, reliably, rapidly, at a low cost and on a global scale.

Through convergence, graphical displays, touchscreens, computer graphics, voice control and human gestures are fast becoming the car's interface, with electronic sensors and algorithms determining much of the driving experience. The software-driven features coming down the road have new auto infotainment apps, but brand new features such as personalization, advanced driver-assistance systems, extensions for car share, region-specific adaptations, car-to-home integration, new vehicle safety options and remote mobile control have become a reality because of connectivity and IoT.

8.1.2 Car to the Cloud

In the past a car's durability and sustainability heavily depended on mechanical parts however as the software complexity overtake the hardware, it is absolutely essential for automotive OEMs and their suppliers to proactively collect on-road car data on operation, performance, cybersecurity, code vulnerability, and human interaction by means of a OTA-based closed-loop data analytics solution.

Car applications and driver interactions can make immense improvements to the marketing, sales, service and product development functions, and help personalize and strengthen customer relationships, and increase revenue by developing various value-added services for better safety. This is all possible by harnessing the car data that is being generated and analyzed. Also, the data helps to diagnose the vehicles and provides preventive to predictive analytics, which can be used to provide information to companies such as insurance agencies, and to determine software updates in the car for its safety and others.

8.1.3 Car Cloud to Everything

The IoT is already gaining significant traction in many areas of the automotive industry. This convergence will optimize process efficiency as actionable intelligence will continually usher novel business opportunities. IoT will accelerate mobile/wireless and cloud innovation due to the increasing use of cloud and the general rise in demand. Advances in Big Data and predictive analytics, will benefit the automotive industry on the whole.

This apart, IoT will get increasingly affordable and practical as the IoT endpoint size and price of sensors fall. Automotive OEMs, dealers and service organizations that are at the forefront of innovative technologies can look forward to great returns, as seen in Fig. 8.2 illustrating the ecosystem for the connected car.

IoT will be leveraging global connectivity, big data and analytics to enable 'Smart Cities'. Through IoT, there will be new capabilities, which will have the ability to remotely monitor, manage and control devices and generate new outputs and information from huge data that is been collected. This data will be extensively utilized and analyzed to bring transformational changes in cities by enhancing infrastructure, getting more cost-efficient municipal services, upgrading public transport and definitely reducing traffic.

Fig. 8.2 IoT applications are key service enabler for the software defined car

8.2 Impact of Convergence

Unexpected business outcomes are set to arise due to rapid IoT-enabled convergence of technologies and platforms. This means that businesses are bound to see lower liability in insurance through driver, vehicle and safety.

This is how, individuals, society and industries will be impacted because of the convergence of automotive and IoT.

8.2.1 Impact to an Individual

Driverless cars will make you pay less for traffic tickets, as it will not be jumping traffic signals or even speeding. Unlike humans, the cars will not be programmed to speed, when running late. Also, the need to own more than one car will be eliminated as this car will do the job for you of picking up your kids from school, when you are at work and will solve other issues, where you need another car or another person to complete the job for you. Just imagine the time that you will save through these cars. It will lower traffic congestion, human stress and will improve the quality of life (QoL) to a great extent. As the industry moves from mechanical to more software and machine learning, it will evidently open up new job opportunities in the software space, where services will be created and we will witness a rise in a new era from digital to data cloud. On the flipside, the convergence will also see individual privacy concern rise, as the autonomous cars will be connected through sensors and other devices.

8.2.2 Impact to the Society

Autonomous cars are capable of determining the best route and warn each other about the conditions ahead. Autonomous cars will allow the vehicles to use the roads more efficiently, thus saving space and time. There are many socio-economic impacts that the society will benefit from with these cars. Cars will be more fuel-efficient and improve safety, the traffic will flow better, and savings will be achieved in professional driving and vehicle maintenance. We will do our bit for the environment by adapting to autonomous cars as this will further lower pollution, because of vehicle-to-vehicle (V2V) communication and more smooth and energy-efficient driving. The city design will drastically change without stressed drivers looking for parking spots selflessly, cars can ease themselves in tiny spaces. End users will lower pollution and carbon foot print by adapting to autonomous cars.

8.2.3 Impact to the Industry

There are other industries that are bound to benefit from the convergence. One of them is the telecommunications industry, since connected vehicles will increasingly communicate over mobile networks, and will generate substantial growth. With connectivity already allowing companies and car clubs to provide access to car as a service, which is expected to be heavily consumed in the years to come. The automotive insurance industry to be slightly disrupted. As safety will improve, premiums will fall to some extent and the number of new cars will start to decline.

8.3 Opportunities and the Future

The increasing scale of data from the vehicle and connected devices represents a remarkable revenue opportunity for the players that can provide insights from data, value from transactions and new innovative services. The raw data that is being generated by on-board sensors, cameras, Lidars etc. will only accelerate as advanced driving assistance system (ADAS) and self-driving cars become a reality—a rate that will reach 1 Gbps by 2020, according to Intel's report. Based on the author's estimate "This data could amount to 5.4 TB daily per vehicle". So imagine if this data was analyzed and used to create vehicle improvements and cost savings. We are currently gathering much less than 1 % of the data generated and car OEMs and Tier-one suppliers know they need to start making use of all this data. Some of these data will be transmitted to the cloud. However, it raises risks for stakeholders as security of data is paramount. Risk of car-hacking is a menace and the industry needs to tackle it at the earliest.

Extraordinary growth of IoT will drive mobile/wireless data, but it is set to lead to scarcity of wireless spectrum and exposure to unprotected data. Privacy of data also remains an issue, as we mentioned earlier. Governments must, therefore, invest heavily in basic technologies such as Internet, mobile, storage and artificial intelligence. Education for the workforce to utilize connected devices to improve overall productivity is also the need of the hour to bring about tangible industry results and face challenges.

America's drivers wasted 6.9 billion hours in traffic in 2014 [3]. This is bound to come down tremendously, and consequently, will shoot up American productivity and scope of a knowledge-based workforce. Benefits of the automotive IoT era will be evident from improved traffic patterns, better fuel economy and enhanced safety. According to the National Highway Traffic and Safety Administration (NHTSA) notice of proposed rulemaking over V2V technology, the aim of this is to ensure that V2V technology is integrated into new vehicles to reduce traffic congestion [1].

8.4 Role of 5G Mobile/Wireless Technology

Mobile technologies such as 3G and 4G LTE along with WiFi, IEEE 802.11n/ac operating in the unlicensed spectrum are the main drivers for making connected car a reality. Automotive OEM in partnership with telecom carriers will need to figure

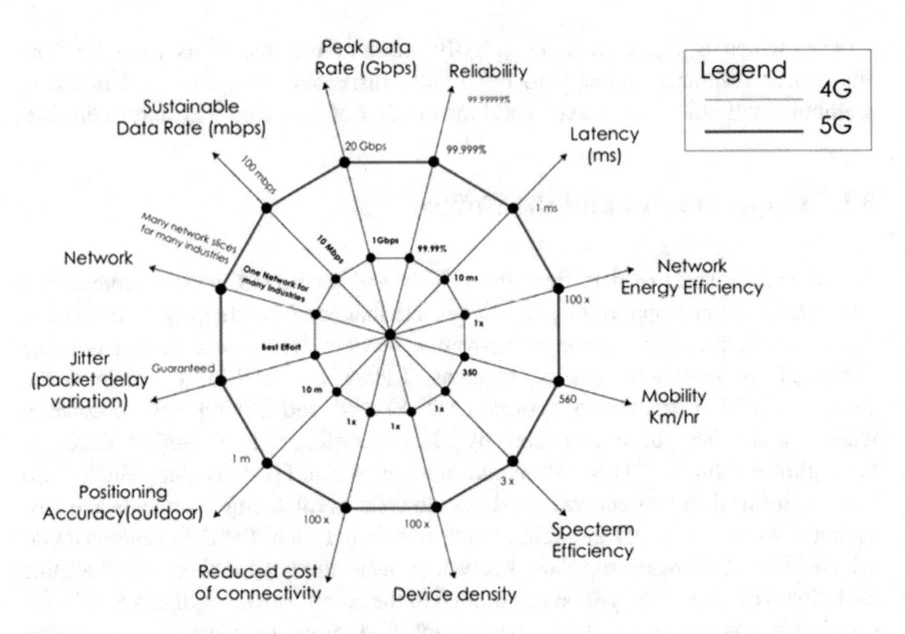

Fig. 8.3 Technical comparison of 5G with 4G systems

out a profitable business model to offer connectivity to all cars (embedded connectivity in car and via tethered smartphone solutions) without any additional charges to consumer, then and only then will the vision for 100 % connected car be achieved by 2020.

With high speed connectivity at the heart of connected car, 5G systems have a significant role to play as the industry undergoes major transformation towards full autonomous car which will require cars to cooperate with each other and with the infrastructure in a secure and reliable manner with high throughput, guaranteed jitter/delivery and reduced latency.

Figure 8.3 shows the mapping of 4G and 5G across the most significant parameters and especially those related to automotive.

5G will cover most of the autonomous driving use cases, however, here below are a few use cases that require attention from the 5G standardization body (see Table 8.1). Figure 8.4 shows the autonomous car requirements overlaid on top of 5G systems.

Table 8.1 Autonomous car use cases not covered By 5G standard

	Use case for autonomous car	Requirements currently outside of 5G standards	
		Position accuracy (Outdoor)	Systems reliability
1	Overtaking vehicle: Bi-directional single lane with slow or heavy duty vehicle in front of it	Under 30 cm	99.99999 % (Seven 9s)
2	Platoon driving: Freeway smooth traffic flow in tight batches	Under 30 cm	In scope of 5G systems
3	Autopilot in urban area: Detection of organic matters like human, pets, etc. and objects like ball, bicycle, etc.	Under 10 cm	99.99999 % (Seven 9s)
4	Vehicle-to-Pedestrian (V2P): Detection of kids, elderly person, physically challenged individuals, etc.	Under 10 cm	In scope of 5G systems
5	Lane-less driving: When lane are not visible (covered with snow) or missing (dirt road)	Under 10 cm	99.99999 % (Seven 9s)

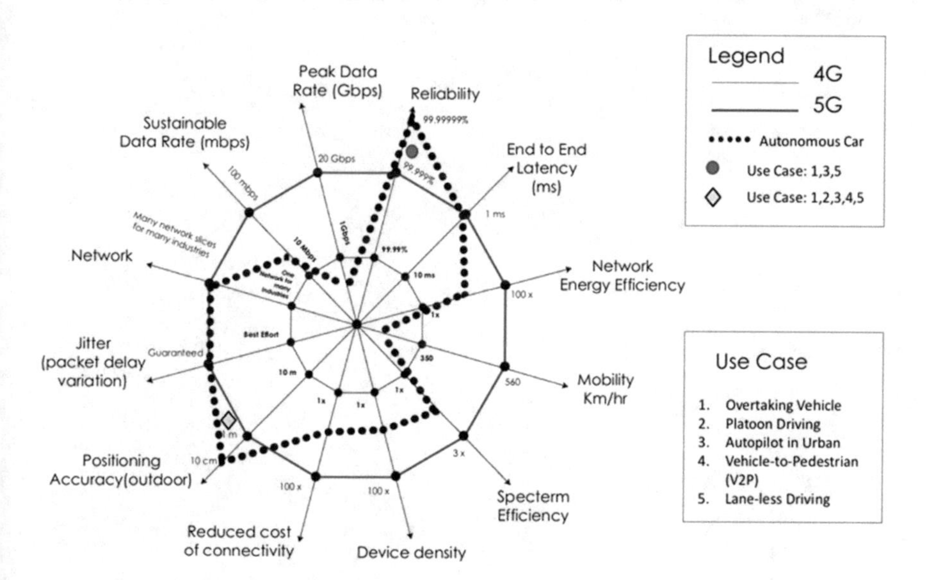

Fig. 8.4 Autonomous use cases mapped on to 5G

8.5 Conclusion

As connected and software-defined cars become common on urban roads, there will be a considerable number of software devices installed in the vehicles. They are tomorrow's supercomputer-on-wheels. IoT will further offer a stimulus to the car industry by connecting and making use of data through the cloud for the 250 million cars just on American roads. Shared driverless cars with 5G connectivity are the future though they may sound like a thing of science fiction today. The remunerations will be increased by businesses across sectors even outside of automotive businesses. IoT will allow everything and everyone to be linked on the go, consequently changing the way we collaborate and consume. The road ahead for the automotive industry is open and lined with opportunities. It's time to shift into top gear and 5G should be at the center of it!

References

1. Musk E, In less than 20 years, owning a car will be like owning a horse. http://www.techinsider.io/elon-musk-owning-a-car-in-20-years-like-owning-a-horse-2015-11
2. This car runs on code. http://spectrum.ieee.org/transportation/systems/this-car-runs-on-code
3. Vehicle-to-Vehicle Communications. http://www.safercar.gov/v2v/index.html
4. Here's How Much Time Americans Waste in Traffic. http://abcnews.go.com/US/time-americans-waste-traffic/story?id=33313765
5. Facts and statistics about M2M. http://www.statista.com/topics/1843/m2m-machine-to-machine/

Chapter 9
Future of Healthcare—Sensor Data-Driven Prognosis

Arpan Pal, Arijit Mukherjee and Swarnava Dey

Abstract Connected Wellness/Healthcare is all about retrieving people's physiological parameters through sensors and performing analysis. Individually, such analytics can help a patient to maintain a wellness regime, or to decide when to see a doctor, and can assist doctors in diagnosis. Collectively, such analytics, if performed over a long time over large set of patients, has the potential to discover new disease diagnostic and treatment protocols. In this chapter, we first discuss how advances in sensing and analytics can take us from a reactive *illness-driven* healthcare system to a proactive *wellness-driven* system. We introduce an IoT driven architecture and discuss how non-invasive, affordable, unobtrusive sensing using mobile phones, wearables and nearables is making physiological and pathological data collection from human body possible in thus far unimaginable ways. We also introduce breakthrough technologies in form of *exosomes* and 3D organ printing that has the potential to disrupt the future healthcare landscape.

Assuming various kinds of physiological and pathological measurements becoming available 24 × 7 from patients, a whole set of analytics opportunities open up for alerting, diagnostics, prognostics and diagnostic rules discovery. We discuss how big data analytics techniques like anomaly detection, stream reasoning, evidence based learning moving towards AI techniques like deep learning and cognitive computing will help in moving from the current *art of diagnostics* to future *science of prognostics*. Finally we take a peek into a hypothetical healthcare system of future where with help of all the above technologies, we describe a day in the life of a person in 2050.

A. Pal (✉) · A. Mukherjee · S. Dey
Innovation Labs, Tata Consultancy Services, Kolkata, India
e-mail: arpan.pal@tcs.com

A. Mukherjee
e-mail: mukeherjee.arijit@tcs.com

S. Dey
e-mail: swarnava.dey@tcs.com

R. Prasad and S. Dixit (eds.), *Wireless World in 2050 and Beyond: A Window into the Future!*, Springer Series in Wireless Technology,
DOI 10.1007/978-3-319-42141-4_9

Healthcare is something that everybody relates to and is always seen as an *imperfect* system where a lot of improvement can be made. This is true for all countries across the globe irrespective of their economic standings. Only the causes of concerns are different for different parts of the world. For example, in developed nations, the problems in healthcare revolve around invasive and costly diagnosis procedures and the fact that some diseases are yet to have a cure. There is also concern that the current healthcare procedure encourages *one size fits all* kind of diagnostic and treatment protocols, where in reality, every individual is different and may require personalized handling. The other big problem that is plaguing developed nations is the considerable rise in ageing population due to availability of better healthcare and increase in longevity. For example, US Government statistics states that there were 44.7 million elderly people in USA which is projected to increase to 98 million in 2060.[1] However, in order to ensure quality-of-life for the elderly citizens, specialized healthcare systems need to be designed that will minimize human-in-loop support systems because at some time there will be dearth of non-elderly people to help and support the huge elderly population.

On the other hand, in developing nations, in addition to the above stated problems, there are some issues very unique to the developing nations only. To start with, there is the capacity problem—there are not enough doctors per patient. It is compounded by the reachability problem—there may still be enough doctors in big cities, but basic primary healthcare is still scarce in the remote rural areas. Finally, while there has been tele-healthcare technologies being developed for delivering remote healthcare to partially take care of the capacity and reachability problem, it needs to be made more affordable in order to benefit masses.

The biggest problem today's healthcare system faces is the fact that it is an *illness-driven* system. All the stakeholders in the current healthcare system—be it doctors, care givers, hospitals, pharmaceutical companies or medical device manufacturers, all benefit when the patient becomes *ill*. The basic business models at force in today's healthcare incentivize the stakeholders of the medical eco system when people are not *well*. This needs a paradigm shift and there is need to strive towards a system that is *wellness-driven* instead of being *illness-driven*. It should incentivise all the healthcare stakeholders to keep patients *healthy*.

Sensing of physiological and pathological parameters to help in diagnosis and deciding on a treatment once the diagnosis is done are backed by extensive and detailed scientific studies. However same cannot be said about the diagnosis processes that lie in-between sensing of signs and symptoms and prescription of treatments. In fact medical diagnosis is still treated by many as an *art* rather than *science*. For this reason, often there are wide variations in diagnosis between doctors and quite frequently there are disagreements between the first and second opinions. It all depends on the experience and knowledgebase of the concerned doctor to arrive at the right diagnosis with no guarantee of repeatability for another patient or another doctor. There is nothing wrong with the doctors—they are

[1]http://www.aoa.acl.gov/Aging_Statistics/index.aspx.

handicapped by the fact that medical science is still an *inexact science* and a vast part of how human body works or reacts is still either unknown or have inadequate scientific evidence. This highlights the need for *evidence-based* diagnosis systems in place of current *rule-based* diagnosis systems which can support personalization and adaptation.

In this chapter, in Sect. 9.1, we first present an Internet-of-Things (IoT) driven remote health monitoring architecture which tries to address the home and elderly care, capacity and reachability issues mentioned above. It also introduces the need for data driven analytics based diagnosis and prognosis system to aid doctors. Such systems will not only help in moving from *illness* to *wellness*-driven models but also help move from *art of diagnostics* to *science of prognostics*. In Sect. 9.2, we introduce affordable, unobtrusive/non-invasive sensing systems that would be the primary means to gather rich, continuous pathophysiological data in order to make the data and evidence driven analytics described in Sect. 9.1 to be feasible. We not only discuss the current state of the art in physiological and pathological sensing, but also outline what is going to happen in immediate future in form of wearable technologies. We also try to provide a glimpse of where human sensing technology may move in far future. Finally in Sect. 9.3, we provide the current status of data driven systems for diagnostic aid in form of decision support systems and rule engines followed by advances happening in statistical processing and machine learning help create better evidence driven diagnostic systems. We also introduce Artificial Intelligence (AI) techniques like deep learning and cognitive computing which has the potential to disrupt the healthcare industry in far future creating the true evidence-driven personalized prognostic science for healthcare.

9.1 IoT Driven Remote Health Monitoring

9.1.1 IoT Architecture for Remote Health Monitoring

The architecture of a typical remote health monitoring system is depicted in Fig. 9.1.

As seen from the figure, there can be medical equipment, mobile phone sensors, wearable, implantable/ingestible or nearable sensors (like optical/thermal camera/RF Imaging devices) which the patient or the concerned people can use in their home to collect their physiological signals like heart rate, blood pressure, ECG, heart sound, blood oxygen etc. Additionally there can be pathological sensing in form of blood glucose, blood analysis etc. All these sensors can be connected over wireless protocols like Bluetooth to a gateway device which most likely can be a mobile phone. There is an emerging medical device interface standard called Continua,[2] which is typically used to interface between medical devices and the

[2]http://www.continuaalliance.org.

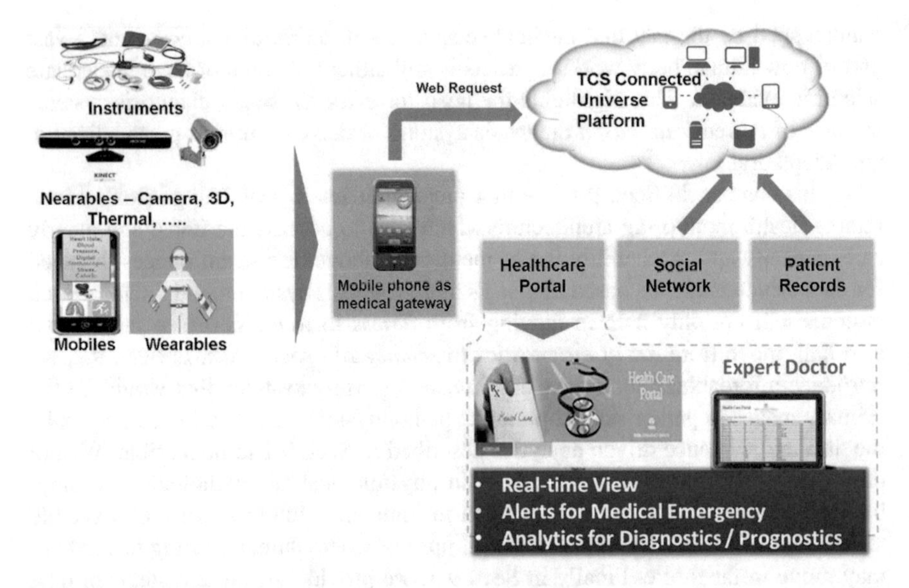

Fig. 9.1 Typical remote health monitoring architecture

gateway. The gateway device is connected over Internet to a Cloud based platform like the TCS Connected Universe Platform (TCUP).[3] In addition to providing the sensor data transport, the platform also provides the necessary patient management, device management and security features.

Sensor data get logged in the cloud storage via the Gateway Interface. Doctors can access the system via secured login to look at the patient data and advice/prescribe accordingly. However this system can be improved in line with the objectives stated in Sect. 9.1 by making the sensing more informative, usable, unobtrusive and affordable, creating automated alert generation and analytics systems for diagnosis/prognosis. Such systems has already being deployed over TCUP platform for use cases like elderly people monitoring, factory worker health monitoring and basic fitness/wellness monitoring [1].

The physiological and pathological sensing is described in detail in Sect. 9.2, first covering the current state-of-the-art, then outlining technologies that are becoming feasible and available in near future and finally introducing possible disruptive technologies. The alert generation sub-system can address the capacity problem stated in introduction. The analytics system can provide the necessary data and evidence-driven prognosis addressing the *migration from illness to wellness* requirement stated in introduction. These are described in detail in Sect. 9.3.

[3] http://www.tcs.com/about/research/Pages/TCS-Connected-Universe-Platform.aspx.

9.2 Disrupting the Status Quo—Affordable, Unobtrusive, Non-invasive Sensing

9.2.1 Connected Medical Devices, Physiological Sensing Using Mobile Phones and Wearable

(a) *Connected Medical Devices*—There are a host of connected standalone medical devices in the market that can measure various physiological parameters like blood pressure, body fat, ECG, blood oxygen saturation etc.[4] There are also connected pathological sensing devices like blood chemistry analyzer, blood hematology analyzer etc.[5] Almost all of them can communicate using Bluetooth or USB and a few of them comply with standards like Continua mentioned earlier for data exchange. However, there are two major issues with such devices. Firstly, individual devices are costly resulting in a high total cost of ownership for a set of measurement devices, which violates the affordability requirement. Secondly, majority of the devices use their own protocols and data formats, thereby making their integration into a larger remote health monitoring system (like one outlined in previous section—Fig. 9.1) a difficult task.

(b) *Mobile Phone based Physiological Sensing*—Most smart-phones today have sensors like accelerometer, gyroscope, magnetometer, microphone and camera. Using these, it is possible to sense some of the basic physiology of human beings that are related to wellness [2]. The advantage of mobile phone based sensing is the affordability—it is a zero cost software add-on to sense as smart phone is already pervasive and available to a large number of population. According to ITU-T,[6] there are 7.085 billion mobile phone subscriptions worldwide in 2015 out of a total world population of 7.3 billion which is a huge penetration even after discounting multiple subscriptions from a single user. Out of it, 42.9 % are smart phones and by 2018, for the first time more than half of mobile phones (51.7 %) will be smart phones.[7] Given this trend, one can expect pervasive coverage of smart phone ownership globally in near future.

Using the accelerometer, gyroscope and magnetometer of the mobile, one can create a basic fitness app on the mobile that can not only count the steps taken or distance covered, but also classify the type of activity (walking/jogging/running) and compute an accurate estimate of the calorie burnt [3]. Using the mobile phone camera, placed on fingertip, it is possible get photoplethysmogram (PPG)

---

[4]http://www.omron-healthcare.com.

[5]http://www.samsung.com/in/business/business-products/healthcare-product.

[6]http://www.itu.int/en/ITU-D/Statistics/Documents/statistics/2015/ITU_Key_2005-2015_ICT_data.xls.

[7]http://www.emarketer.com/Article/2-Billion-Consumers-Worldwide-Smartphones-by-2016/1011694.

waveforms by using image and signal processing techniques on the fingertip video. From the PPG, it is not only possible to measure the heart rate and respiratory rate [4], but it is also possible to measure the blood pressure [5]. People have also reported measuring blood oxygen saturation (SpO2) from the same PPG signal obtained from mobile phone camera; however there are associated challenges which need to be solved [6]. Using the mobile phone microphone, it is possible to get the heart sound and use it as a digital stethoscope [7]. Not only that, people have reported doing Spirometry based Lung Function Test using mobile phone microphones [8]. These research works suggest that a whole lot of physiological parameters that spans from general fitness to health of heart and health of lung, can be measured using mobile phone sensors.

(c) *Wearable based Physiological Sensing*—Wearable devices are expected to lead the next device revolution. Their market penetration is increasing and they are becoming more and more affordable. The main advantage of wearable devices is their capability to do 24 × 7 monitoring in an unobtrusive manner. There are already a number of wearable devices available in the market that can measure the basic fitness parameters like steps taken, distance walked etc.[8] Such devices use accelerometers within the gyroscope to measure the required parameters. However, wearable devices are becoming more and more capable and it is possible to not only sense PPG signal from a wrist wearable[9] but also get ECG waveforms from it.[10] From the PPG signals, the usual heart rate, respiratory rate and blood pressure can be measured.

Wearable devices are still plagued by battery consumption issues and hence most of the available devices today use the mobile phone as a gateway for transmission using low power protocols like Bluetooth Low Energy (BLE)[11] instead of connecting to the internet directly via a cellular connection. As the battery technology improves and energy harvesting becomes technically feasible, there will be more proliferation of wearable devices for 24 × 7 physiological monitoring.

9.2.2 In Near Future—Wearable Pathological Sensing, Nearable Devices

(a) *Wearable Pathological Sensing*—While the wearable technologies for physiological sensing are maturing, significant progress has also been made in wearable pathological sensing. The current pathology systems are invasive in the sense that blood or other in-body fluid samples need to be collected.

[8]https://www.fitbit.com.

[9]http://www.shimmersensing.com/.

[10]https://www.empatica.com/.

[11]http://www.bluetooth.com/Pages/low-energy-tech-info.aspx.

The future generation pathological sensing either uses easily available outer body fluid like sweat or teardrops. There have been reports of early success in measuring blood glucose from tear drops [9]. Google has announced their intent to build a smart diabetic sensing contact lens using the above technology.[12] There are also reports of doing body nutrient analysis using tear drops.[13] Scientists have also reported measurement of ethanol, drug, ion and metal content from body sweat [10]. Startup organizations have now been founded who aim to build wearable sensors that can sense and analyze body sweat for measuring hydration, fluid loss and electrolytic imbalance.[14] In addition to wearable, such sensing can also be done through smart ingestible pills.[15]

(b) *Nearable Devices*—By nearable devices we mean camera and RF sensing devices which can be placed near a person to gather information about physiology. The main advantage of nearable sensing is its unobtrusiveness because of which no external device or internal implant need to be placed on a person's body. There has been early success reported in recent work from MIT which can sense heart rate from normal optical camera signals using 3D image processing tracking micro movements in our face due to blood pumping [11].

Another group at MIT is also working on detecting heart rate and breathing rate from wireless RF signals by modelling the RF interference of human body [12]. These technologies, once mature would be a low-cost, unobtrusive, constraint-free option to monitor health of multiple patients without putting anything on patient's body.

9.2.3 Future Vision—Exosomes, 3D Organ Printing

In addition to the above, there are some exciting, path-breaking work happening, which if successful will disrupt the way we look at healthcare. A few of them are outlined below.

(a) *Exosomes—"Secreted vesicles known as exosomes were first discovered nearly 30 years ago. But, considered little more than garbage cans whose job was to discard unwanted cellular components, these small vesicles remained little studied for the next decade. Over the past few years, however, evidence has begun to accumulate that these dumpsters also act as messengers, actually conveying information to distant tissues. Exosomes contain cell-specific*

[12]http://www.healthline.com/health-news/diabetes-google-develops-glucose-monitoring-contact-lens-012314.

[13]http://www.superiorideas.org/projects/infant-teardrop.

[14]http://www.wired.com/2014/11/sweat-sensors/.

[15]http://www.marsdd.com/news-and-insights/ingestibles-smart-pills-revolutionize-healthcare/.

payloads of proteins, lipids, and genetic material that are transported to other cells, where they alter function and physiology”.[16] It is widely believed that the exosomes carry information about our current body condition and health. By analyzing exosome data, it may become possible in distant future to diagnose diseases and predict illnesses, thereby paving the way for true wellness driven healthcare. Since exosomes are secreted vesicles, they have the possibility to be sensed via sweat on the skin.

(b) *3D organ printing*—Replacement of defective organs in the body like liver, heart, lung etc. has been an established process. However, availability of matching donors and acceptability of the transplanted organ by the body remain big issues. Scientists have started working on creating synthetic organs using additive 3D printing via tissue engineering [13]. However this research has a long way to go before feasibility of such systems can be thought of. One complex challenge in building 3D printed organs is the creation of nano-vascular structures within the organ through which blood can flow.

9.3 Disrupting the Status Quo—Predictive Analytics

Analysis of healthcare data for better healthcare processes has been one of the most talked about areas for more than two decades and have resulted in the development of a number of solutions ranging from analyses of patients' pathophysiological symptoms for better diagnosis to analyses of hospital data for better healthcare management processes. Though, in the perception of the majority, use of computers in medicine has typically implied assistance in diagnosis of diseases, and in this section, our focus will remain in that area of diagnosis and prognosis in healthcare using various analytical means.

The first requirement for medical data analytics is successful gathering and storage of data. Electronic Medical Records (EMR) are emerging as the standard to store the data. As more and more healthcare organizations are storing medical records in electronic format, the availability of hospital data and EMRs are also increasing. However, neither the hospital datasets nor the EMR data can directly be used by analytics algorithms due to the inherent problems associated with those. Clinical datasets like MIMIC [14] are quite large but very unclean, containing noises from the bio medical instruments, different artifacts and other error. There have been significant research efforts for automated cleaning of such data [15–17].

Automated EMR data extraction requires extensive use of Natural Language Processing (NLP) techniques to identify and aggregate similar information from words, statements and concepts in the documents. As detailed in the report by Pivovarov and Elhadad [18], it is also important to perform temporal reasoning over the records, find relative ordering of events and represent that in the data to be

[16]http://www.the-scientist.com/?articles.view/articleNo/30793/title/Exosome-Explosion/.

analyzed. To represent extracted information from EMRs, it is necessary to use existing clinical knowledge, available in form of ontologies and terminologies including ICD codes,[17] UMLS[18] and RxNorm.[19]

9.3.1 *Evolution of Diagnostic Systems—From Rule-Based CDSS to Learning Systems*

During the past two decades, significant effort has been put towards the development of *Clinical Decision Support Systems* (*CDSS*) based on prominent data mining techniques, such as decision trees, associative rule mining etc. CDSS have traditionally been developed following two different approaches. The *knowledge based* systems were built by integrating expert doctors' knowledge about symptoms, disease conditions and drug administrations. These systems employed reasoning methods on the stored knowledgebase and provided decisions/recommendations at the point of care. One of the famous examples of such a CDSS is MYCIN [19], which was developed around the mid-eighties as an expert system employing *Artificial Intelligence* in medicine. The knowledgebase comprised of association rules and facts on drug interactions. A query from any user would initiate a backward traversal of these association rule chains, creating implicit rules which would be validated using stored data. The major contribution of MYCIN was the application of deductive reasoning on medical knowledge using a goal-directed approach, where the system started with a goal statement and through backward inference, found the data that could establish that goal.

The knowledge based CDSS were reliable but non-scalable like any other expert system, due to the difficulty of getting expert knowledge. However, not only the non-scalability issue was a major area of concern, the knowledge based CDSS could not gather sustained interest due to access to limited amount of data (from experts/manual inputs), preventing those systems from finding out novel insights from data, hitherto unknown to the experts/doctors themselves. Thus to overcome these limitations, focus of CDSS related research shifted towards creating *machine learning* based systems that would automatically mine knowledge from evidence.

Over the past decade, major research efforts were directed towards medical knowledge mining from medical datasets and records. Different machine learning algorithms were used to extract association rules among diseases and patient parameters. Disease/symptoms, disease/drugs appear as frequent sets in the data, revealed by applying suitable learning algorithms. One of the most well-known such algorithm is Association Rule Mining (ARM), detailed discussion of which is

[17] http://apps.who.int/classifications/icd10/browse/2016/en.

[18] https://www.nlm.nih.gov/research/umls.

[19] https://www.nlm.nih.gov/research/umls/rxnorm.

beyond the scope of this chapter, but the works presented by Ji and Deng [20], Leung et al. [21] and more recently González [22] deal with ARM in a generalized context. In medical diagnostics, work presented in [23] finds the factors influencing heart disease; [24] performs ARM in image data; [25] applies association determination in context of diabetes prevention.

Diagnostic systems based on rules and reasoning engines have a requirement of understanding the context, as the interpretation of data is largely dependent on the context, especially for sensor-driven systems as is the central theme of this article [26]. Kara and Dragoi, in [27], have argued that context is volatile in nature, and to make diagnostic systems context-aware, decisions and actions must happen dynamically, periodically and on-demand. Thus, it is imperative that in order to become successful, knowledge-based expert systems must also offer contextual interpretation of pathophysiological data.

Knowledge mining CDSS have great possibilities as those systems can find insights from data, which even an expert doctor might find helpful in diagnosing a disease. Thus these systems are more suitable as an interface that can help clinicians to take decisions quickly and effectively, rather that providing answers to disease detection/drug administration related questions. However, as noted before, scalability has always been a concern as modeling the knowledge is a difficult problem which may require involvement of several other techniques such as Information Retrieval, Natural Language Processing, using which medical texts from journals, books, trials etc. can be processed and interpreted.

A major drawback of the knowledge-based CDSS is that these are not trusted by many clinicians. To scrutinize the mined rules, extensive factual knowledge is a mandatory requirement; and in the domain of healthcare, making this knowledge machine-understandable is extremely difficult, especially because the diagnostic procedures are dependent on personal experiences of clinicians.

9.3.2 Emergence of Statistical/Linear Learning Models Over Neural Networks

The difficulties with CDSS with respect to the incorporation of prior knowledge and rules led researchers in the field of machine learning to explore other possibilities. During early nineties, *Artificial Neural Networks* (*ANNs*) were an exciting field of research and were thought to be an option to explore in the area of biomedicine. Neural networks were shown to be capable of approximately solving any complex non-linear equation by learning through multiple iterations over a set of data. For problems related to medical imaging or medical signal analysis, such an approach seemed to be a perfect match. However, the major problem with this approach was the relative lack of understanding of the internal working of the neural nets, and a major section of the current learning systems have been developed using linear models of supervised and unsupervised feature extraction and classification of patterns using a training dataset, mostly due to the ease of analysis and physical interpretation of such models. Further, non-linear neural nets were computationally expensive and a relative lack of good training data led to an almost parallel emergence of the linear and statistical models as a viable alternative. Sajda in [28] and Kononenko in [29], explain in details the issues faced with neural nets and the benefits of the linear and statistical models. Such models have been used fairly successfully in many exemplary bodies of work, such as those highlighted in [30–33].

The learning systems in this category function in two phases: (i) *feature extraction*, or the method of extracting useful features which often are more informative from the raw signal, and training of the system using the feature set and (ii) *classification*, or the means to infer the underlying physical organization of the elements within the data/feature set. During the past decade, many useful techniques have been developed for identifying a useful set of features, such as, *singular value decomposition* (SVD), or *principal component analysis* (PCA) and more recently, *mutual information coefficient* (MIC) which is particularly useful in identifying non-obvious relationships between individual features in a dataset. However, all these methods require an already identified set of features on which they will operate in order to select the most relevant ones. This often appears as a problem, especially while processing sensor signals, as there can be numerous features within a given signal, and there is no clear mapping as to which feature may give the best estimate of a physiological parameter, for example, in the case of deriving the blood pressure of a person from the *photoplethysmograph* (PPG) obtained at his/her fingertips. Further, the *curse of dimensionality*[20] at times increases the problem manifolds.

Researchers have used well-known classification/clustering techniques such as *k-Means* and *support vector machines* (SVM) for the final classification of biomedical data, but, some inherent characteristics of biomedical data led to a wide-spread adoption of Bayesian inference models. Sajda, in [28], analyzes this development as well, and attributes this to the associated uncertainty in biomedical data. Biomedical data, especially data obtained from sensors and devices are often noisy and incomplete; an extremely pertinent example can be the *arterial blood pressure* (ABP) signal obtained from patients admitted to ICUs, where the signal contains many anomalies that are attributed to sudden muscle movements or seizures, or certain electrical leads suddenly becoming loose. Statistical methods to detect such anomalies are in existence and often used before feeding the dataset to the learning system. In addition to the inherent noise, often measurements are inconsistent with prior knowledge. Bayesian network theory allows creation of generative models offering a generic approach especially for image and signal processing which can deal with the uncertainty and variability within the data/measurements.

[20]https://en.wikipedia.org/wiki/Curse_of_dimensionality.

9.3.3 Future Vision—Cognitive Computing and Artificial Intelligence

In recent years, there has been a change in the computing landscape because of the emergence of concepts such as the Cloud, big data platforms, GPUs, and more importantly, the re-emergence of neural networks as a possible computing model—all of which together have led to a collective belief that modeling the human brain may not be just a fictional notion any more. The field of *Artificial Intelligence* has gone through spectacular breakthroughs, such as machine-learning-based checkers game (1950), simulation of neurons (1954), development of Eliza—the first NLP program (1966), Shakey—the AI robot (1972) to one of the most talked about victories of machines over humans in 1997 when Deep Blue defeated the reigning world chess champion, Gary Kasparov. However, more recent developments in this area which began with the victory of IBM's Watson over the human champion in a natural language quiz show called *Jeopardy*! in 2011, the advances in the Google Brain[21] project since 2011 and the public release of AI-based digital personal assistants such as Cortana (Microsoft 2014) *who* learns about the habits of the human user, have been cornerstones in the most talked-about research area that attempts to compute like human brains do—*Cognitive Computing*. And it is widely believed that these technologies can cause an immense disruption in the area of healthcare diagnosis and prognosis.

It has already been noted earlier that identifying and extracting the most relevant features from a dataset or signal is a difficult task, and depends heavily on the knowledge about the signal and domain. Recent advances in deep-layered neural networks, commonly known as *deep learning systems*, indicate that it may be possible for the network to learn the features at different levels from the raw input data/signal on its own. Traditional learning systems required labeled/annotated data to train itself for correctly classifying the input, a fact which appeared difficult in healthcare systems, because of the manual nature of the annotation task. Researchers in the area of deep learning have shown that unsupervised feature learning can indeed be applicable in a variety of tasks [34, 35] and new learning algorithms such as *Restricted Boltzman Machine* (RBM) [36] and its variations, *Stacked auto-encoders* [37] etc. have resulted in much greater accuracy in identifying patterns within the input data/signals.

The successes of deep neural networks in pattern detection problems (especially within images and speech) have inspired researchers to use similar approaches in healthcare diagnosis. Längkvist in [38] have used RBMs to classify sleep stage data with a greater accuracy than conventional learning systems based on Gaussian mixture and hidden Markov models. Wulsin et al. [39] have attempted to model electroencephalography (EEG) signals using Deep Belief Networks composed of RBMs for fast classification and anomaly measurement in a semi-supervised

[21]http://googleblog.blogspot.com/2012/06/using-large-scale-brain-simulations-for.html.

manner (where a small percentage of the training dataset was labelled). Längkvist worked on estimating the bacteria content by using unsupervised deep learning techniques on signals from electronic nose in [40]. Similarly, Mirowski et al. used time-delay neural networks to discover patterns for epilieptic seizures from EEG signals [41]. A considerably large number of researches on diagnosis and prognosis in healthcare using deep neural nets over time-series data [42] or signals from sensors can been found in the AI landscape in recent times indicating that researchers have found this area to have potentials for successful futuristic healthcare systems.

On the other side of the spectrum, lies *DeepQA*—a complementary approach of using knowledge bases to answer questions in natural language in variety of subjects, such as demonstrated by Watson in 2011 [43]. The challenge in this area is mostly associated with modelling the knowledge effectively using information retrieval from a plethora of materials, of heterogeneous type, structure and language, implying successful application of natural language processing techniques so that it can be searched in real-time to find the answer to a question, and assign a measure of confidence to the answer. Kelly in [44], outlines a range of application scenarios for such DeepQA systems, which can act as clinicians' assistants by analysing pathophysiological data, radiology reports, gene sequences, answering questions on symptoms and drugs etc.

Our belief is that a successful futuristic healthcare diagnosis and prognosis system will be a hierarchical system of three or more tiers—with the first tier being an anomaly detection layer to generate a clean set of signals which will act as inputs to the diagnosis and prognosis system; the second tier in our belief, could be a deep neural net, meant for analysing input data to discover features followed by classification or prediction with a certain probability score; a third tier could be a reasoning layer which uses rules specified in a knowledgebase created using information retrieval and NLP techniques to ascertain the score given by the previous tier. For example, a deep neural net, by analysing a multivariate input of different physiological signals, such as heart rate, blood pressure, oxygen saturation may predict a 60 % chance of heart related anomalies; at the reasoning level, further personalised inputs such as the patient's lifestyle parameters (smoker or not), family history of diabetes and heart anomalies etc. can be considered, which may increase the score to a much higher level if the patient has a history of smoking and/or there is a family history of heart anomalies, or may decrease the score if the person leads a healthy lifestyle.

9.4 Conclusion—A Day in the Life of a Patient in 2050

In this chapter we have outlined how future healthcare systems are shaping up considering the current research efforts in Internet of Things, connected devices and sensors, signal processing, machine learning and AI. We now try to look ahead in the future, and attempt to sketch the outline of a healthcare system that may exist in

2050, based on what we have outlined so far in this article. The characters, places, structures mentioned in this section are fictional and have been borrowed from George RR Martin's epic, *A Game of Thrones*.[22]

"Almost seven hundred feet high it stood, three times the height of the tallest tower in the stronghold it sheltered. His uncle said the top was wide enough for a dozen armored knights to ride abreast. The gaunt outlines of huge catapults and monstrous wooden cranes stood sentry up there, like the skeletons of great birds, and among them walked men in black as small as ants."[23] As dawn broke on the Wall, Jon Snow, the Lord Commander of the Night's Watch, was dreaming about the day he saw the colossal structure for the first time. He did not have a good night's sleep—it has been nearly twenty years on the Wall since he came here at the age of eighteen—twenty years of vigilance, hard rangings in the frozen forests beyond the Walls, bloody battles with terrors beyond the Wall to keep the men in the South safe; for the Night's Watch were the swords in the darkness; they were the shields that guard the realms of men.

As the dawn broke, Maester Aemon's face appeared in view and his voice ringed within the room—"Good Morning, Jon! It seems from your EEG signals that you had trouble sleeping last night; you even dreamed of your first view of the Wall and felt the same scare as you did then." Maester Aemon used to be the maester on the Wall—the scholar, the healer, and the advisor. But he has passed away, most old maesters have. The new maesters have recently created a medical system where they have added the wisdom of the old, which together with the processing of the data collected by all the devices on a person's body and around, can diagnose a person's health without any maester actually seeing the person. They have even personalized the image and voice that comes out of the system while communicating with someone, by selecting an appearance and voice once trusted by the person, like Maester Aemon for Jon Snow. Jon Snow was wearing a little ring that captured his body signals which now appeared on the displays scattered across the room. "You are still dragging your left leg and the ankle joint has a distortion of one degree. I assume you are doing all the exercises shown to you, Jon?"—Aemon's 3D hologram continued while looking at the skeletal display of Jon via the tiny camera mounted on the sidewall. Jon nodded and while he was about to send for his squire asking for breakfast, the android squire wheeled into the room with a tray containing some food and infusions on his hand. "I have taken the liberty to prepare your breakfast Jon," Aemon continued, "by matching your pathophysiological profile with that of your biological father, Late Lord Eddard Stark of Winterfell, it appears that you have a high risk of cardiac problems and unless you start controlling your diet and take the infusions specially created matching your gene profile, your ranging days will be short in numbers."

After breakfast and his daily physio routines in front of the camera satisfying Maester Aemon, Jon came out on the portico to take a look at the ranging party

[22]https://en.wikipedia.org/wiki/A_Game_of_Thrones.

[23]http://awoiaf.westeros.org/index.php/Wall.

assembled at the gate of Castle Black. Twenty-five fresh looking members of the Night's Watch, each having gone through similar routines such as Jon Snow with the medical system (via the impression of maesters once trusted by the individuals) performing the daily check-up on each of them, gathered around on their mounts. Each of them had a device, either a ring, or a pendant, or likewise, continuously gathering the physiological signals and transmitting back to the data processing nodes for real-time analysis of their physical and cognitive alertness based on which the First Ranger can customize his team and strategies, in real-time. The gate opened and the ranging party advanced northwards, onto the icy paths of the frozen forest in search of any possible threats to the wall. Twenty years after the battle of the five kingdoms, the skies beyond the Wall seemed silent and the young members of Night's Watch alert and vigilant; for they are the swords in the darkness; they are the watchers on the walls; they are the shields that guard the realms of men.

Acknowledgments We thank all our fellow scientists and engineers at Innovation Lab, TCS for their help, inputs and observations.

References

1. Ghose A, Sinha P, Bhaumik C, Sinha A, Agrawal A, Choudhury AD (2013) UbiHeld: ubiquitous healthcare monitoring system for elderly and chronic patients. In: Proceedings of the 2013 ACM conference on pervasive and ubiquitous computing adjunct publication. ACM, pp 1255–1264
2. Khan WZ, Xiang Y, Aalsalem MY, Arshad Q (2013) Mobile phone sensing systems: a survey. IEEE Commun Surv Tutorials 15(1):402–427
3. Chandel V, Choudhury AD, Ghose A, Bhaumik C (2014) AcTrak-unobtrusive activity detection and step counting using smartphones. In: Mobile and ubiquitous systems: computing, networking, and services. Springer, pp 447–459
4. Pal, A, Sinha A, Choudhury AD, Chattopadyay T, Visvanathan A (2013) A robust heart rate detection using smart-phone video. In: Proceedings of the 3rd ACM MobiHoc workshop on pervasive wireless healthcare. ACM, pp 43–48
5. Banerjee R, Ghose A, Choudhury AD, Sinha A, Pal A (2015) Noise cleaning and Gaussian modeling of smart phone photoplethysmogram to improve blood pressure estimation. In: IEEE international conference on acoustics, speech and signal processing (ICASSP). IEEE, pp 967–971
6. Karlen W, Lim J, Ansermino JM, Dumont G, Scheffer C (2012) Design challenges for camera oximetry on a mobile phone. In: Annual international conference of the IEEE engineering in medicine and biology society (EMBC). IEEE, pp 2448–2451
7. Brusco M, Nazeran H (2004) Digital phonocardiography: a PDA-based approach. In: 26th annual international conference of the IEEE engineering in medicine and biology society, IEMBS'04, vol 1. IEEE, pp 2299–2302
8. Larson, EC, Goel M, Boriello G, Heltshe S, Rosenfeld M, Patel SN (2012) SpiroSmart: using a microphone to measure lung function on a mobile phone. In: Proceedings of the 2012 ACM conference on ubiquitous computing. ACM, pp 280–289
9. Yan Q, Peng B, Su G, Cohan BE, Major TC, Meyerhoff ME (2011) Measurement of tear glucose levels with amperometric glucose biosensor/capillary tube configuration. Anal Chem 83(21):8341–8346. doi:10.1021/ac201700c. 12 Oct 2011

10. Jadoon S, Karim S, Akram MR, Kalsoom Khan A, Zia MA, Siddiqi AR, Murtaza G (2015) Recent developments in sweat analysis and its applications. Int J Anal Chem 2015. Article ID 164974, http://dx.doi.org/10.1155/2015/164974
11. Wu HY, Rubinstein M, Shih E, Guttag J, Durand F, Freeman W (2012) Eulerian video magnification for revealing subtle changes in the world. ACM Trans Graph 31, 4. Article 65. doi:http://dx.doi.org/10.1145/2185520.2185561
12. Adib F, Mao H, Kabelac Z, Katabi D, Miller RC (2015) Smart homes that monitor breathing and heart rate. In: Proceedings of the 33rd annual ACM conference on human factors in computing systems. ACM, pp 837–846
13. Merceron TK, Burt M, Seol Y-J, Kang H-W, Lee SJ, Yoo JJ, Atala A (2015) A 3D bioprinted complex structure for engineering the muscle tendon unit. Biofabrication 7(3):035003
14. Mohammed S, Villarroel M, Reisner AT, Clifford G, Lehman L-W, Moody G, Heldt T, Kyaw TH, Moody B, Mark RG (2011) Multiparameter intelligent monitoring in intensive care II (MIMIC-II): a public-access intensive care unit database 39:952–960. doi:10.1097/CCM.0b013e31820a92c6
15. Sweeney KT, Ward TE, McLoone SF (2012) Artifact removal in physiological signals—practices and possibilities, information technology in biomedicine. IEEE Trans 16(3): 488–500. doi:10.1109/TITB.2012.2188536. Date of Publication: 22 Feb 2012
16. Akhtar MT, Mitsuhashi W, James CJ (2012) Employing spatially constrained ICA and wavelet denoising, for automatic removal of artifacts from multichannel EEG data. Sig Process 92(2): 401–416. ISSN 0165-1684, http://dx.doi.org/10.1016/j.sigpro.2011.08.005
17. Yorkey TJ (1997) Method and apparatus for removing artifact and noise from pulse oximetry. U.S. Patent No. 5,645,060. 8 Jul 1997
18. Pivovarov R, Elhadad N (2015) Automated methods for the summarization of electronic health records. J Am Med Inform Assoc 1–12. doi:10.1093/jamia/ocv032
19. Buchanan BG, Shortliffe EH (1984) Rule based expert systems: the MYCIN experiments of the Stanford heuristic programming project. Addison-Wesley, Reading, MA. ISBN 978-0-201-10172-0
20. Ji C-R, Deng Z-H (2007) Mining frequent ordered patterns without candidate generation. In: Fourth international conference on fuzzy systems and knowledge discovery, FSKD 2007, vol 1. IEEE
21. He H-T, Zhang S-L (2007) A new method for incremental updating frequent patterns mining. In: Second international conference on innovative computing, information and control, ICICIC'07. IEEE
22. Rodríguez-González AY et al (2013) Mining frequent patterns and association rules using similarities. Expert Syst Appl 40(17):6823–6836
23. Nahar J et al (2013) Association rule mining to detect factors which contribute to heart disease in males and females. Expert Syst Appl 40(4):1086–1093
24. Olukunle A, Ehikioya S (2002) A fast algorithm for mining association rules in medical image data. In: Canadian conference on electrical and computer engineering, IEEE CCECE 2002, vol 2, pp 1181–1187. doi:10.1109/CCECE.2002.1013116
25. Carnethon MR, Prineas RJ, Temprosa M et al (2006) Diabetes prevention program research group. The association among autonomic nervous system function, incident diabetes, and intervention arm in the diabetes prevention program. Diabetes Care 29:914–919
26. Mukherjee A, Pal A, Misra P (2012) Data analytics in ubiquitous sensor-based health information systems. In: 6th international conference on next generation mobile applications, services and technologies (NGMAST). IEEE
27. Kara N, Dragoi OA (2007) Reasoning with contextual data in telehealth applications. In: 3rd international conference on wireless and mobile computing, networking and communications
28. Sajda P (2006) Machine learning for detection and diagnosis of disease. Ann Rev Biomed Eng 8:537–565
29. Kononenko I (2001) Machine learning for medical diagnosis: history, state of the art and perspective. Artif Intell Med 23(1): 89–109. ISSN 0933-3657, http://www.sciencedirect.com/science/article/pii/S093336570100077X

30. Shankaracharya OD, Samanta S, Vidyarthi AS (2010) Computational intelligence in early diabetes diagnosis: a review. Rev Diabet Stud 2010 Winter 7(4):252–262
31. Zhao X, Cheung LW (2007) Kernel-imbedded gaussian processes for disease classification using microarray gene expression data. BMC Bioinform 8. 28 Feb 2007
32. Han X (2007) Cancer molecular pattern discovery by subspace consensus kernel classification. Comput Syst Bioinform Conf 6:55–65
33. Statnikov A, Aliferis CF, Tsamardinos I, Hardin D, Levy S (2005) A comprehensive evaluation of multicategory classification methods for microarray gene expression cancer diagnosis. Bioinformatics 21(5):631–643
34. Coates A, Ng AY, Lee H (2011) An analysis of single-layer networks in unsupervised feature learning. In: International conference on artificial intelligence and statistics
35. Ranzato MA et al (2007) Unsupervised learning of invariant feature hierarchies with applications to object recognition. In: IEEE conference on computer vision and pattern recognition, CVPR'07. IEEE
36. Salakhutdinov R, Mnih A, Hinton G (2007) Restricted Boltzmann machines for collaborative filtering. In: Proceedings of the 24th international conference on Machine learning. ACM
37. Vincent P et al (2010) Stacked denoising autoencoders: learning useful representations in a deep network with a local denoising criterion. J Mach Learn Res 11:3371–3408
38. Längkvist M, Karlsson L, Loutfi A (2012) Sleep stage classification using unsupervised feature learning. Adv Artif Neural Syst 2012:5
39. Wulsin DF et al (2011) Modeling electroencephalography waveforms with semi-supervised deep belief nets: fast classification and anomaly measurement. J Neural Eng 8(3):036015
40. Längkvist M, Loutfi A (2011) Unsupervised feature learning for electronic nose data applied to bacteria identification in blood. NIPS 2011 workshop on deep learning and unsupervised feature learning
41. Mirowski PW, Madhavan D, LeCun Y (2007) Time-delay neural networks and independent component analysis for eeg-based prediction of epileptic seizures propagation. In: Proceedings of the national conference on artificial intelligence, vol 22, No. 2. AAAI Press, MIT Press, Menlo Park, CA, Cambridge, MA, London, 1999, 2007
42. Längkvist M, Karlsson L, Loutfi A (2014) A review of unsupervised feature learning and deep learning for time-series modeling. Pattern Recogn Lett 42: 11–24. (http://www.sciencedirect.com/science/article/pii/S0167865514000221). 1 Jun 2014
43. Ferrucci D, Brown E, Chu-Carroll J, Fan J, Gondek D, Kalyanpur AA, Lally A et al (2010) Building Watson: an overview of the DeepQA project. AI Mag 31(3):59–79
44. Kelly III J, Hamm S (2013) Smart machines: IBM's Watson and the era of cognitive computing. Columbia University Press

Chapter 10
Big Data and Internet of Things—Challenges and Opportunities for Accelerated Business Development Beyond 2050

George Suciu

Abstract Big Data (BD) and Internet of Things (IoT) are considered to generate major businesses and impact on the workforce market until 2050 and beyond. However, there are barriers that are still slowing down the success of BD and IoT applications, especially due to the lack of standardization and challenges of different players to cooperate. On the other side, access to large scale cloud computing testbeds has become a commodity and huge opportunities can be exploited in domains such as agriculture, automotive, renewable energy, health, and smart cities. The goal is to present business development ideas and steps that a young and smart entrepreneur should take to start and accelerate a successful career in BD/IoT, scaling-up beyond 2050.

The world is faced with challenges in three big dimensions of sustainable development—economic, social and environmental. More than 1 billion people are still living in extreme poverty, and income inequality within and among many countries has been rising; at the same time, unsustainable consumption and production patterns have resulted in huge economic and social costs and may endanger life on the planet. Achieving sustainable development will require global actions to deliver on the legitimate aspiration towards further economic and social progress, requiring growth and employment, and at the same time strengthening environmental protection.

Sustainable development will need to be inclusive and take special care of the needs of the poorest and most vulnerable. Strategies need to be ambitious, action-oriented and collaborative, and to adapt to different levels of development. They will need to systemically change consumption and production patterns, and might entail, inter alia, significant price corrections, encourage the preservation of natural endowments, reduce inequality and strengthen economic governance [1].

G. Suciu (✉)
BEIA Consulting, București, Romania
e-mail: george@beia.ro

R. Prasad and S. Dixit (eds.), *Wireless World in 2050 and Beyond: A Window into the Future!*, Springer Series in Wireless Technology,
DOI 10.1007/978-3-319-42141-4_10

The advances in information technology have witnessed great progress on healthcare technologies in various domains nowadays. However, these new technologies have also made healthcare data not only much bigger but also much more difficult to handle and process. Moreover, because the data are created from a variety of devices within a short time span, the characteristics of these data are that they are stored in different formats and created quickly, which can, to a large extent, be regarded as a big data problem [2].

Big Data is an area of information technology indicating the collection and management of large amounts of data, which cannot be effectively and efficiently handled with classic data management techniques such as RDBMS. The need for handling such large datasets comes from the possibility of deriving analytic information by finding correlations among data. Big data refers to the practice of collection and processing of very large data sets and associated systems and algorithms used to analyze these massive datasets [3].

The Internet of Things (IoT) is the network of physical objects or "things" embedded with electronics, software, sensors, and network connectivity, which enables these objects to collect and exchange data [4]. IoT allows objects to be sensed and controlled remotely across existing network infrastructure, creating opportunities for more direct integration between the physical world and computer-based systems, and resulting in improved efficiency, accuracy and economic benefit; when IoT is augmented with sensors and actuators, the technology becomes an instance of the more general class of cyber-physical systems (CPS), which also encompasses technologies such as smart grids, smart homes, intelligent transportation and smart cities [5].

Each thing is uniquely identifiable through its embedded computing system but is able to interoperate within the existing Internet infrastructure. Experts estimate that the IoT will consist of almost 50 billion objects by 2020 [6]. Others consider that IoT will be a rapidly growing segment, but have moderate forecasts of 26 billion connected devices by 2020 [5].

We believe that connectivity is the key enabler for making the Internet of Things happen and cellular has an important role to play. The ability of intelligent devices to perceive and respond to the environment around them makes them highly valuable for complex, automated decision making in a broad range of industries.

The paper is organized as follows: Sect. 10.1 describes the identified challenges, Sect. 10.2 presents the main business opportunities. Section 10.3 presents the 2050-2059 timeline contents and Sect. 10.4 concludes the paper.

10.1 Challenges for Big Data and IoT

Challenges that are still slowing down the success of IoT applications/ecosystems from SME point of view (and of course, which solutions would work) are: access to cloud—becoming more easy to get, access to funding—Startup programs encouraging IoT topics and Access to telecom operators—promoted using contests and portals.

Standardization is another challenge identified, as different protocols and standards existing and under development (ZigBee, XMPP, oneM2M, IEEE P2413, etc.) and the approach will be to use standardized M2M with IPv6, allowing 100 addresses for every atom on the face of the earth. The mass distribution of video programming over IP networks promises a richer experience for viewers, with widely predicted increases in interactivity, choice, personalization, and the ability to micro pay for a la carte programming. Whereas broadcasting was licensed, controlled and regulated tightly by national governments (or even owned as a monopoly service), video-over-IP will be delivered by international market mechanisms with both relatively minimal direct legal restraint and little direct government strategic intervention. Standardizing video delivery to produce network economies of scale and scope will require international corporate coordination between the converging industries of broadcasting and video production, wired and wireless telecommunications, and computer hard- and software derived data communications [7].

Ray Kurzweil has constructed this logarithmic chart that combines 15 unrelated lists of key historic events since the Big Bang 15 billion years ago (Fig. 10.1) [8]. The exact selection of events is less important than the undeniable fact that the intervals between such independently selected events are shrinking exponentially.

Kurzweil wrote with great confidence, in 2005, that the Singularity would arrive in 2045. One thing I find about Kurzweil is that he usually predicts the nature of an event very accurately, but overestimates the rate of progress by 50 %. Part of this is

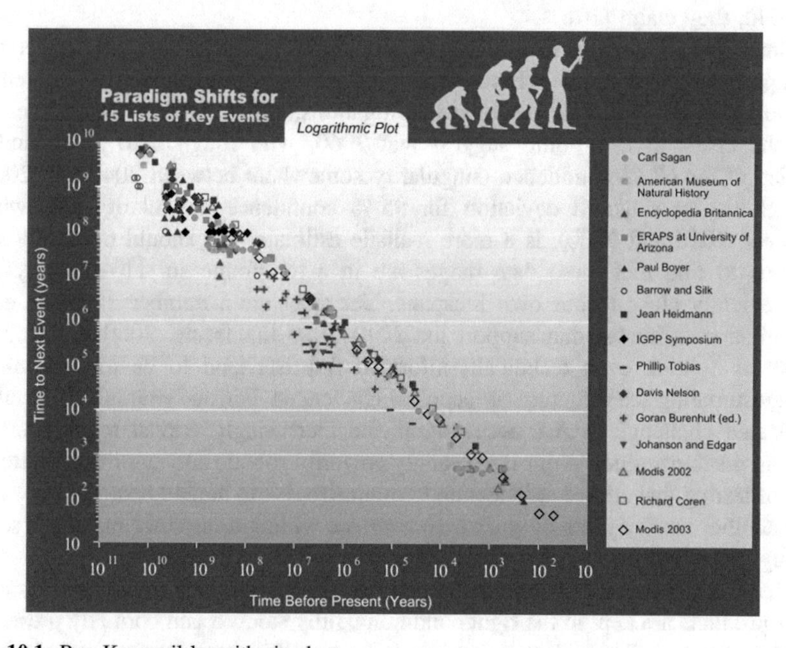

Fig. 10.1 Ray Kurzweil logarithmic chart

because he insists that computer power per dollar doubles every year, when it actually doubles every 18 months, which results in every other date he predicts to be distorted as a downstream byproduct of this figure. Another part of this is that Kurzweil, born in 1948, is taking extreme measures to extend his lifespan, and quite possibly may have an expectation of living until 100 but not necessarily beyond that. A Singularity in 2045 would be before his century mark, but herein lies a lesson for us all. Those who have a positive expectation of what the Singularity will bring tend to have a subconscious bias towards estimating it to happen within their expected lifetimes. We have to be watchful enough to not let this bias influence us. So when Kurzweil says that the Singularity will be 40 years from 2005, we can apply the discount to estimate that it will be 60 years from 2005, or in 2065 [9].

By 2050, a completely new type of human will evolve as a result of radical new technology, behavior, and natural selection. This is according to Last [10], a researcher at the Global Brain Institute, who claims mankind is undergoing a major 'evolutionary transition'. By 2040, cabs will be driven by Google robots, shops will become showrooms for online outlets and call centers will be staffed by intelligent droids. That's the scenario depicted in recent research which suggests robots could be taking over our lives and jobs in less than 30 years. The competition for work caused by a rise in the robots population will see us heading to surgeons for 'additional processing power for our brains', they claim. We may also be requesting bionic implants for our hands that will make us able to perform tasks as fast as any machine. Futurologists, commissioned by global job search website xpatjobs.com, say workers will have less job security and will work more unsociable hours. Those who take these risks and innovate with their own bodies will be the biggest earners in 2040, they claim [10].

Smart [11] is a brilliant futurist with a distinctly different view on accelerating change from Ray Kurzweil, but he has produced very little visible new content in the last 5 years. From his personal investigations, which are as much a guess as anyone else's at this point, suggest that 2060, with rough ±20 year standard deviation for 68 % confidence (singularity somewhere between 2040 and 2080), and ±40 year standard deviation for 95 % confidence (singularity somewhere between 2020 and 2100), is a more realistic estimate. We should be highly suspicious to find that these date ranges are in a timeframe so self-servingly and conveniently close to our own lifespans. Yet there are a number of useful early quantitative estimates that support the 2040–2080 timeframe. 2060 is later than most, as I think even technically-informed futurists tend to be too optimistic, underestimating the difficulty of technical challenges. For one example of an often neglected challenge to A.I. acceleration, the increasingly critical requirement of adding machine ethics to our increasingly powerful robotic and automated systems, and of testing that ethical architecture to minimize risk to human beings, may *easily* add another twenty years of work before we are willing to produce machines smart enough to reliably and safely direct their own self-improvement.

Retail faces an ever changing competitive landscape and for FMCG's ensuring their products are kept in the right conditions, fully stocked and correctly presented can be a huge challenge. Retail companies spend millions—sometimes hundreds of millions—of euros into retail display cabinets for better product placement and improved quality. Cabinets are typically provided to stores free of charge with the promise that they will increase sales to cover the investment. The challenge is that after the cabinets are delivered, the owner has very little visibility and control over these assets.

Key issues:

- Owner doesn't typically know where the cabinets are at any given time due to unauthorized moves or even theft—resulting in reported 5–15 % annual loss of assets;
- Cabinets are often switched off to save electricity, resulting in poor product quality e.g. warm beverage—this has a huge impact on product sales;
- Store manager uses the cabinet for competitor products, making the whole investment pointless for the owner—purity is a significant challenge especially for beverage companies;
- Even when assets are working, the ability to measure when stock needs replenishing is limited to 'after the event' reporting. This often means the assets sit idle until new stock arrives;
- The impact of the above is that the return on investment for retail cabinets is far from optimal. Luckily, there is a solution.

The private motor insurance market has faced a number of long-term challenges. Relatively short purchasing cycles mean that motor insurance policies typically come up for renewal every six to 12 months. The industry has a high churn rate, and competition is fierce for the more profitable low-risk drivers in the more favorable geographies. The removing of gender as an underwriting factor (EC Gender Directive) is beginning to have a major impact on premiums. Profitability has been the main focus for many providers, with premiums being too low and claims inflation going through the roof. There has been a concerted effort by the industry to rectify this by pushing through major increases in premium levels. Consumers are now faced with a multitude of choices, with an array of insurance now available at their fingertips. It is essential for insurance providers to differentiate themselves in both marketing and sales processes from the competition if they want to attract new and also retain existing business.

Connecting enterprise assets, such as printers, vending machines, lifts or even heart monitors, and receiving live updates on their operational status presents a huge opportunity. This enables process automation, new usage-based business models and completely new revenue streams. The challenge with connecting assets is managing the vast amounts of data which can become overwhelming making it impossible to drive value from the data. Assets also require effective, automated remote monitoring and management to avoid costly site visits. Faced with increased competition businesses are demanding new revenue streams and process efficiency, and without a comprehensive management solution in place, businesses are exposed to unnecessary complexity, operational inefficiency and additional cost.

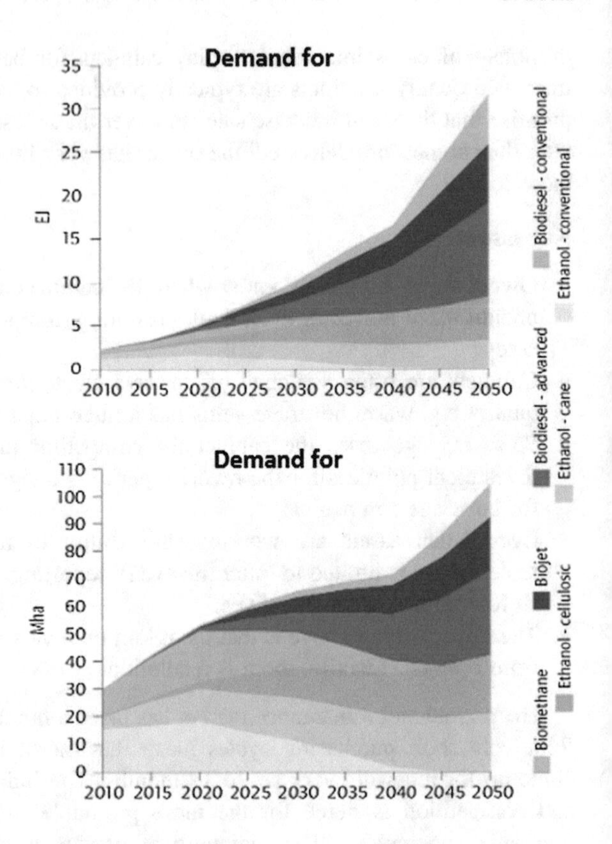

Fig. 10.2 Demand for biofuels (*top*) and resulting demand for land (*bottom*)

Biofuels also will be a huge source of competition for diminishing resources available for food production. According to the International Energy Agency, biofuel production will see an 800 % increase between now and 2050. While much of that biofuel will come from nonfood crops and second-generation production techniques such as cellulosic ethanol, most of the current supply of biofuels and production in the near term will provide direct competition with resources used to grow crops for human consumption and feed for livestock. Projected growth in biofuel demand also is expected to require more than triple the land currently used for production, as shown in the bottom graph of Fig. 10.2, further intensifying competition between food crops and biofuel crops [12].

10.2 Opportunities for Big Data and IoT

In 2050, as highways, shipping and communication have gone completely digital—all kinds of devices connected to the web, big data and artificial intelligence to "make people more aware of their world and their own behavior" [13].

By 2020, the market for the IoT industry alone is expected to rise to $7.1 trillion, and by the year 2050, technology experts predict that there will be 50 billion devices that are connected to the Internet (Cisco)—considered too conservative (other predict 100 billion). And, clearly all humans will eventually be able to access the Internet, wherever they are. All 7 billion of us (and then 8 billion, and then 9 billion of us) all interacting seamlessly and intimately on the Internet in new digital worlds of our own making [14].

IoT can enable and accelerate many new business and service opportunities, as well as revenue generation. After the “wearables” revolution, the “intimates” and “internals” revolution will probably not be far behind. Telehealth is reality in your own home, as big data, telehealth, and the cloud come together 70 % renewables by 2050? It’s doable with ‘Internet of Energy’: opportunities presented by Internet of Things-based communications software and storage technologies are what ultimately will drive change and European Commission (EC) has adopted the Communication “Energy Roadmap 2050” [15].

In 2010, 50 % population lived in cities and in 2050 is expected that 70 % will be city dwellers. 6.4 billion people and 2/3 of buildings needed by 2050 are not yet built—account of half of global energy consumption. For this, we propose smart cities as an opportunity. Smart City projects are designed for reducing emissions, increasing the use of renewable energy sources in urban areas, such as wind turbines, solar panels and biogas, and also reduce water consumption, also prevent pollution of water sources [16].

Furthermore, drones and robots are expected to generate a market of $67 Billion up to 2050 [17], with opportunities such as Smart Dust, Elevator to the moon, etc. but the ancient art of origami being more than just child’s play; its principles are being used to meet a wide range of modern engineering challenges—from how to fold up a tennis court-sized solar array for use in space, to reducing the amount of waste in a tube of toothpaste.

Agriculture to feed 9 billion is another opportunity for IoT and Big Data, and involves that the farmers and big data companies to invest in precision agriculture, using sensors, drones, satellites and GPS tracking systems, farmers using only one type of precision technologies increased their yield by 16 % and cut down water use by 50 % and rapid payback using these technologies—a 15 % savings on seed, fertilizer, and chemicals [17].

3D printing is becoming increasingly capable and affordable. We envision a future world where interactive devices can be printed rather than assembled; a world where a device with active components is created as a single object, rather than a case enclosing circuit boards and individually assembled parts. This capability has tremendous potential for rapid high fidelity prototyping, and eventually for production of customized devices tailored to individual needs and/or specific tasks. With these capabilities we envision it will be possible to design highly functional devices in a digital editor—importing components from a library of interactive elements, positioning and customizing them, then pushing ‘print’ to have them realized in physical form [18].

Moreover, 3D sensing will present new opportunities, such as biosensors consisting of a bio-recognition component, biotransducer component, and electronic system which include a signal amplifier, processor, and display. Transducers and electronics can be combined, e.g., in CMOS-based microsensor systems [19]. The recognition component, often called a bioreceptor, uses biomolecules from organisms or receptors modeled after biological systems to interact with analyze of interest. This interaction is measured by the biotransducer which outputs a measurable signal proportional to the presence of the target analyze in the sample. The general aim of the design of a biosensor is to enable quick, convenient testing at the point of concern or care where the sample was procured [20].

Another opportunity can be the 3D vision for assets tracking. Efficiently managing critical business assets such as tools, supplies, equipment, fleet or products poses a significant challenge and can be a highly labor-intensive and costly process. Additionally, the theft, loss, breakdown or delay in delivery of assets can have huge financial implications—loss of working time, delays in production, the expense of hiring in alternative equipment and increased insurance premiums. With increasing competition demanding greater flexibility and process efficiency, the inability to track and recover assets due to poor process controls and lack of automation can have a significant impact. Businesses need to look at how they can do more with less—how they can automate manual processes, increase efficiencies, reduce theft and loss and optimize uptime and utilization of assets.

Technology will play a central role in driving change. Some of the value being created in 2050 will derive from wholly unanticipated breakthroughs but many of the technologies that will transform manufacturing, such as additive manufacturing, are already established or clearly emerging. Table 10.1 summarizes some of the most important pervasive and secondary technologies including ICT, sensors, advanced materials and robotics. When integrated into future products and networks, these will collectively facilitate fundamental shifts in how products are designed, made, offered and ultimately used by consumers. Mass personalization of low-cost products, on demand: The historic split between cheap mass produced products creating value from economies of scale and more expensive customized products will be reduced across a wide range of product types. Technologies such as additive manufacturing, new materials, computer-controlled tools, biotechnology, and green chemistry will enable wholly new forms of personalization. Direct customer input to design will increasingly enable companies to produce customized products with the shorter cycle-times and lower costs associated with standardization and mass production. The producer and the customer will share in the new value created [21].

Distributed production: We will see a transformation in the nature of production itself, driven by trends such as new forms of modelling and additive manufacturing through to nanotechnologies and advanced robotics. The factories of the future will be more varied, and more distributed than those of today. The production landscape will include capital intensive super factories producing complex products; reconfigurable units integrated with the fluid requirements of their supply chain partners; and local, mobile and domestic production sites for some products. Urban sites will

Table 10.1 Important pervasive and secondary technologies for future manufacturing activities

Pervasive technology	Likely future impacts
Information and communications technology (ICT)	Modelling and simulation integrated into all design processes, together with virtual reality tools will allow complex products and processes to be assessed and optimized, with analysis of new data streams
Sensors	The integration of sensors into networks of technology, such as products connected to the internet, will revolutionize manufacturing. New data streams from products will become available to support new services, enable self-checking inventories and products which self-diagnose faults before failure, and reduced energy usage
Advanced and functional materials	New materials, in which the UK has strong capabilities, will penetrate the mass market and will include reactive nanoparticles, lightweight composites, self-healing materials, carbon nanotubes, biomaterials and 'intelligent' materials providing user feedback
Biotechnology	The range of biotechnology products is likely to increase, with greater use of fields of biology by industry. There is potential for new disease treatment strategies, bedside manufacturing of personalized drugs, personalized organ fabrication, wide availability of engineered leather and meat, and sustainable production of fuel and chemicals
Sustainable/green technologies	These will be used to reduce the resources used in production including energy and water, produce clean energy technologies, and deliver improved environmental performance of products. Minimizing the use of hazardous substances
Secondary technology	
Big data and knowledge based automation	These will be important in the on-going automation of many tasks that formerly required people. In addition, the volume and detail of information captured by businesses and the rise of multimedia, social medial and the internet of things will fuel future increases in data, allowing firms to understand customer preferences and personalize products
Internet of things	There is potential for major impacts in terms of business optimization, resource management, energy minimization, and remote healthcare. In factory and process environments, virtually everything is expected to be connected via central networks. Increasingly, new products will have embedded sensors and become autonomous
Advanced and autonomous robotics	Advances are likely to make many routine manufacturing operations obsolete, including healthcare and surgery, food preparation and cleaning activities. Autonomous and near-autonomous vehicles will boost the development of computer vision, sensors including radar and GPS, and remote control algorithms. 3D measurement and vision will be able to adapt to conditions, and track human gestures

(continued)

Table 10.1 (continued)

Pervasive technology	Likely future impacts
Additive manufacturing (also known as 3D printing)	This is expected to have a profound impact on the way manufacturers make almost any product. It will become an essential 'tool' allowing designs to be optimized to reduce waste; products to be made as light as possible; inventories of spare parts to be reduced; greater flexibility in the location of manufacturing; products to be personalized to consumers; consumers to make some of their own products; and products to be made with new graded composition and bespoke properties
Cloud computing	Computerized manufacturing execution systems (MES) will work increasingly in real time to enable the control of multiple elements of the production process. Opportunities will be created for enhanced productivity, supply chain management, resource and material planning and customer relationship management
Mobile internet	Smart phones and similar devices are positioned to become ubiquitous, general purpose tools for managing supply chains, assets, maintenance and production. They will allow functions such as directed advertising, remote healthcare and personalization of products. Linked technologies include battery technology, low energy displays, user interfaces, nano-miniaturisation of electronics, and plastic electronics

become common as factories reduce their environmental impacts. The factory of the future may be at the bedside, in the home, in the field, in the office and on the battlefield.

Digitized manufacturing value chains: Pervasive computing, advanced software and sensor technologies have much further to go in transforming value chains. They will improve customer relationship management, process control, product verification, logistics, product traceability and safety systems. They will enable greater design freedom through the uses of simulation, and they will create new ways to bring customers into design and suppliers into complex production processes.

One context where robots surely have a future is in space. In the second part of this century the whole solar system will be explored by flotillas of miniaturized robots. And, on a larger scale, robotic fabricators may build vast lightweight structures floating in space (solar energy collectors, for instance), perhaps mining raw materials from asteroids. These robotic advances will erode the practical case for human spaceflight. Nonetheless, we hope people will follow the robots, though it will be as risk-seeking adventurers rather than for practical goals. The most promising developments are spearheaded by private companies. For instance SpaceX [22], led by Elon Musk, who also makes Tesla electric cars, has launched unmanned payloads and docked with the Space Station. He hopes soon to offer orbital flights to paying customers. Wealthy adventurers are already signing up for a week-long trip round the far side of the Moon—voyaging further from Earth than anyone has been before (but avoiding the greater challenge of a Moon landing and blast-off). I'm told they've sold a ticket for the second flight but not for the first flight. We should surely cheer on these private enterprise efforts in space—they can tolerate higher risks than a western government could impose on publicly-funded civilians, and thereby cut costs.

The two key operational benefits across multiple industrial sectors of IoT are [22]:

- Real-time analytics are performed autonomously by intelligent machines and devices in IoT.
- Adaptive analytics: by enabling large-scale, real-time decision making to improve the behavior of devices and systems, adaptive analytics actually makes smart systems smarter.

Another Acceleration opportunities are:

- Supplier attention—Internet of Things developer tools and products are now available.
- Technological advances—Some of the semiconductor components that are central to most Internet of Things applications are showing much more functionality at lower prices.
- Increasing demand—Demand for the first generation of Internet of Things products (fitness bands, smart watches, and smart thermostats, for instance) will increase as component technologies evolve and their costs decline.
- Emerging standards—Over the past two years, semiconductor players have joined forces with hardware, networking, and software companies, and with a number of industry associations and academic consortiums, to develop formal and informal standards for Internet of Things applications.

10.3 2050-2059 Timeline Contents

In this chapter we will present the 2050-2059 timeline contents [23], analyzing the effects of Big Data and Internet of Things. In 2050 is expected that world population will be at 10 billion people, 6 billion Internet user, with 1 billion geolocated social media observations daily, while computers will outnumber humans 10:1, being 250,000 times faster than in 2016, as SSD storage will exceed 500 billion GB.

In 2050:

- **Humanity is at a crossroads**—The world of 2050 is a world of contrasts and paradoxes. On the one hand, science and technology have continued to advance in response to emerging crises, challenges and opportunities. This has created radical transformations in genetics, nanotechnology, biotechnology and related fields.
- **Nearly half of the Amazon rainforest has been deforested**—Droughts caused by global warming have further contributed to the decline, with many areas of jungle being turned into parched scrubland. By 2050, nearly 2.7 million sq km have been deforested. As a result, over 30 billion tons of carbon have been added to the atmosphere.
- **Wildfires have tripled in some regions**—Rising global temperatures are creating drier conditions for vegetation—producing larger and more frequent wildfires. With so much extra burning, air quality and visibility in the western United States is being significantly altered. There has been a 40 % rise in organic carbon aerosols and other smoke particles. These irritate the lungs, but are especially dangerous to people who have trouble breathing as a result of asthma and other chronic conditions. Southern Europe is also badly affected—especially Greece, which has been ravaged in recent decades.
- **Traditional wine industries have been severely altered by climate change**—By 2050, many of the world's most famous wine-producing areas have been rendered unsuitable for traditional grape growing and winemaking, with climate change having severely impacted land use, agricultural production and species ranges. In response to the crisis, many traditional vineyards have shifted to higher elevations with cooler conditions—putting pressure on upland ecosystems, as water and vegetation are converted for human use. Others have made use of genetic engineering, or indoor growing methods such as vertical farming. Geoengineering efforts are also getting underway, but have yet to be successful on a global basis.
- **Fish body size has declined by nearly a quarter**—By far the greatest impact from global warming has been in the seas and oceans, where changes in heat content, oxygen levels and other biogeochemical properties have devastated marine ecosystems. About half of this shrinkage has come from changes in distribution and abundance, the remainder from changes in physiology. The tropics have been the worst affected regions.
- **Hi-tech, intelligent buildings are revolutionizing the urban landscape**—In the first half of the 21st century, a soaring urban population posed serious problems for the environment, health and infrastructure of many cities. Amid worsening climate change and resource depletion, urban regions were forced to either evolve, or die off. Countless cities failed to make this transition in time, and went the way of Detroit, many being abandoned and left to decay, or subject to intense military control and martial law. In those that survived, a new generation of buildings and infrastructure emerged based on these rapidly changing social and environmental needs.

- **Smaller, safer, hi-tech automobiles**—More people than ever before are choosing to live and work alone, while the number of children per couple has also declined, two additional factors which have led to these lighter, more compact vehicles, a large percentage of which carry just one or two passengers. The vast majority of cars in the developed world are now computer-controlled, while traffic flow and other road management issues are handled by advanced networks of AI. The resulting fall in congestion has boosted some economies by tens of billions of dollars.
- **Major advances in air travel comfort**—Commercial airliners of 2050 are safer, quieter and cleaner than those of earlier decades. The vast majority are based on some form of renewable energy. In addition, travel times have greatly improved. Hypersonic engines, which entered use in 2033, have seen further development, aided by the rapid growth of artificial intelligence and the resulting advances in computer-automated design evolution. It is now possible to reach anywhere on the planet in under 2.5 h.
- **Continent-wide "supergrids" provide much of the world's energy needs**—The need for reliable, clean, cost-effective energy has led to the creation of electrical "supergrids" across much of the world. Long distance transmission technology has seen major advances over the decades. Each country is connected to the grid using high-voltage direct current (HVDC) transmission, instead of traditional alternating current (AC) lines. This results in far greater efficiency, since DC lines have much lower electrical losses over long distances.
- **China completes the largest water diversion project in history**—The South-North Water Transfer Project—proposed almost a century ago—is finally completed in China this year at a cost of over $60 billion. This becomes the largest project of its kind ever undertaken, stretching thousands of kilometers across the country.

In 2052:

- **An interstellar radio message arrives at Gliese 777**—The Yevpatoria RT-70, located at the Center for Deep Space Communications in Ukraine, was among the largest radio telescopes in the world, with a 70 m antenna diameter. On 1st July 1999, it beamed a noise-resistant message named "Cosmic Call 1" into space. This was sent towards Gliese 777, a yellow subgiant star, 52 light-years away in the constellation of Cygnus. At least two extrasolar planets were known to be present in this system. In April 2051, the message arrives at its destination, for any potential alien civilizations to hear and decode it.
- **Britain holds its centennial national exhibition**—A centennial national exhibition is held in the UK, in keeping with the precedent set by the Great Exhibition of 1851 and the 1951 Festival of Britain. The opening ceremony is attended by King William V, now aged 69.

In 2053:

- **Moore's Law reaches stunning new levels**—Due to Moore's Law, the average desktop computer now has the raw processing power equivalent to all of the human brains on Earth combined. There is no longer a clear distinction between human and machine intelligence. Entities of astonishing realism and interactivity are widespread. Many are in fact *merging* with human intelligence, as the trend towards brain-computer links increases.
- **Genetically engineered "designer babies" for the rich**—The ability to manipulate DNA has come a long way since its discovery in 1953. A century on, wealthy parents now have the option of creating "perfect" babies in the laboratory. This is done by picking and choosing their best hereditary traits. Gender, height, skin, hair and eye color—along with hundreds of other characteristics—can be programmed into the embryo prior to birth. The embryo is then grown in an artificial uterus. The most advanced (and controversial) techniques involve manipulating the brain to improve the child's intelligence, behavior and personality. Many conservative and religious groups decry what they see as the commercialization of the human body.

In 2054:

- **Rainfall intensity has increased by 20 %**—Dramatic increases in surface runoff, peak river flows and flash flooding are being experienced around the world—exacerbating soil erosion and putting huge pressure on drainage and sewage systems. This additional rainfall is a particular problem in the tropics and poor regions with insufficient infrastructure or flood defenses.

In 2055:

- **Spaceflight has taken a leap forward**—Environmental catastrophes, overpopulation, war and other crises have made humanity painfully aware of the limitations on its home planet. As a result of this, spaceflight has advanced considerably since the beginning of the century. The number of known planets beyond our Solar System—about 800 in 2012—has grown to 13 million by 2055. Thousands of these bodies have been observed in the habitable "Goldilocks" zones of their respective star systems, including a number of Earth-like planets with liquid water oceans and active hydrological cycles. The possibility of finding alien life expands greatly during this time, as does the hope of achieving contact with intelligent civilizations.
- **The vast majority of countries have achieved democracy**—Climate change is now having a significant impact on regional stability, particularly in Africa and the Middle East, where concerns over scarcity of resources have created conditions allowing dictators and authoritarian governments to make a comeback. In any case, a number of cultures are simply more compatible with monarchies, theocracies and autocracies at the present time. These parochial nations will remain undemocratic for some time to come.

- **Global population is reaching a plateau**—The global population is stabilizing at between 9 and 10 billion. Most of the recent growth has occurred in the developing world. However, better education along with improved access to contraception, family planning and other birth control methods is now markedly reducing the number of children per couple. Information technology has played a major role in boosting literacy levels and spreading knowledge to the world's poor.
- **Traditional media have fragmented and diversified**—By the mid-2050s, traditional Western news corporations no longer exist. News gathering, analysis and distribution has fragmented—shifting to millions of creative individuals, bloggers, citizen journalists and small-scale enterprises. These work cooperatively and seamlessly, utilizing a "global commons" of instantly shared knowledge and freely available resources. This includes information retrieval not only from cyberspace, but also in the real world; embedded in everything from webcams and personal digital devices, to orbiting satellites, robots, vehicles, roads, street lamps, buildings, stadia and other public places. Even people themselves have become a part of this collection process. Bionic eye implants (for example) can relay data and footage on the spot, in real time, from those willing to participate.

In 2056:

- **Global average temperatures have risen by 3 °C**—Global warming has begun to race out of control with temperatures fed by increasingly strong feedback mechanisms. Melting permafrost in the Arctic is now releasing vast amounts of methane—a greenhouse gas more than 70 times stronger than CO_2. Plants are decaying faster in the warmer climate, while the oceans are liberating ever greater quantities of dissolved CO_2. The Earth is now the hottest it has been since the mid-Pliocene, over 3 million years ago, and there are permanent El Niño conditions—resulting in widespread, extreme weather events in regions around the world. Severe droughts, torrential flooding, hurricanes and other disturbances are now a constant feature on the news. Southeast Asia, the Middle East and Africa are the places most affected. Developing countries dependent on agriculture and fishing—especially those bordering the Pacific Ocean—are particularly badly hit.
- **Fully synthetic humans are becoming technically feasible**—By 2056, the number of cells that can be synthesized in a single organism is reaching almost 100 trillion: equal to the total number in the human body.

In 2057:

- **Computers reach another milestone**—Computers are becoming so powerful that many high-level tasks in business and government are being handed over directly to them. For years, software had lagged behind hardware in development, which impeded the spread of AI, but this is no longer the case. Nevertheless, it is becoming obvious to everyone by now that machines are quite literally taking over the world.

- **Handheld MRI scanners**—A new generation of machines began to evolve, based on supersensitive atomic magnetometers, detecting the tiniest magnetic fields. These replaced the enormous doughnut-shaped magnets used in the past. By the late 2050s, MRI scans have become as quick and easy as taking a photograph, with a hundredfold decrease in cost.

In 2058:

- **A billion human brains can be simulated in real time**—The late 20th and early 21st centuries witnessed orders of magnitude increases in computer power and data storage. Each new generation of chips was smaller and more energy efficient than the last, resulting in ever larger and more complex applications. This trend was known as Moore's Law and it led to the gradual emergence of artificial intelligence, combined with brain simulations down to the level of single neurons.
- **The Beatles' music catalogue enters the public domain**—Copyright law has remained largely unchanged since 2019. Accordingly, the Beatles' songs from 1962 are entered into the public domain, 96 years after the band's first single.
- **A radio telescope is built on the Moon**—The telescope is 100 m wide and located on the Moon's far side, giving it a stable platform with slow rotation rate, beyond the interference of Earth's atmosphere and cluttered radio background. This provides astronomical images with a clarity unmatched by any observatory on Earth or in space.

In 2059:

- **The end of the oil age**—For most of the 20th century, prospectors discovered far more oil than industrial societies could consume. This was an era of cheap and plentiful energy, which saw huge growth in the world's economy and population. By the late 2050s, the end of the 200-year oil age is approaching, with the final dregs being extracted in the Middle East.
- **Mars has a permanent human presence by now**—By the end of this decade, a permanent team of scientists is present on Mars. This comprises a highly international mix of people. The first civilian tourist has also arrived. Travel to Mars was made cheaper and faster thanks to nuclear pulse propulsion, cutting journey times from six months to just a few weeks.

10.4 Conclusion

Big Data (BD) and Internet of Things (IoT) are considered to generate major businesses and impact on the workforce market until 2050 and beyond. However, there are barriers that are still slowing down the success of BD and IoT applications, especially due to the lack of standardization and challenges of different players to cooperate.

On the other side, access to large scale cloud computing testbeds has become a commodity and huge opportunities can be exploited in domains such as agriculture, automotive, renewable energy, health, and smart cities.

Robots could be taking over our lives and jobs: by 2040, cabs will be driven by Google robots, shops will become showrooms for online outlets and call centers will be staffed by intelligent droids.

H2050 Opportunities and challenges were identified in all major domains and a turning point defined as the Singularity is expected to happen around H2050—our intelligence will become increasingly nonbiological and trillions of times more powerful than it is today—the dawning of a new civilization that will enable us to transcend our biological limitations and amplify our creativity, the so-called Internet of Everything.

Hyper-Acceleration is expected to take place beyond H2050—competition for work caused by a rise in the robots population which will see humans heading to surgeons for 'additional processing power for our brains' and requesting for bionic implants for our hands that will make us able to perform tasks as fast as any machine.

References

1. World Economic and Social Survey (2013) Sustainable development challenges. United Nations publication
2. Suciu G, Fratu O, Vulpe A, Butca C (2015) Cloud and big data acceleration platform for innovation in environmental industry. In: 2015 IEEE international Black Sea conference on in communications and networking (BlackSeaCom) pp 132–136, 18–21 May 2015
3. Suciu G, Vulpe A, Craciunescu R, Butca C, Suciu V (2015) Big data fusion for eHealth and Ambient Assisted Living Cloud Applications. In: 2015 IEEE International Black Sea conference on in communications and networking (BlackSeaCom) pp 102–106, 18–21 May 2015
4. Harvard Business School (2014) A Harvard Business review Analytic Services Report internet of things: Science fiction or business fact?
5. Vermesan O, Friess P (2013) Internet of things: converging technologies for smart environments and integrated ecosystems. Denmark
6. Evans D (2011). The internet of things: how the next evolution of the internet is changing everything. Cisco. Retrieved 4 Sept 2015
7. Marsden C The challenges of standardisation: towards the next generation internet. Available at SSRN: http://ssrn.com/abstract=287634 or http://dx.doi.org/10.2139/ssrn.287634
8. Kurzweil Ray (2015) The singularity is near. Viking Books, New York
9. The futurist 20 Aug 2009 http://www.singularity2050.com/2009/08/
10. Last C (2014) Global brain and the future of human society. World Future Review 2014, pp 1–22
11. Smart J (2007) Futurist, "Types of Futures Thinking", www.accelerationwatch.com. Retrieved 2 March 2007
12. Dutia SG (2014) Agtech: Challenges And opportunities for sustainable growth. Ewing Marion Kauffman Foundation
13. Posel S (2014) 2050: How the internet will monitor our daily lives. The US Independent, 12 June 2014

14. Spencer L (2014) Internet of Things market to hit $7.1 trillion by 2020: IDC, June 5, 2014—http://www.zdnet.com/article/internet-of-things-market-to-hit-7-1-trillion-by-2020-idc/
15. McKinsey Global Institute (2013) Disruptive technologies: advances that will transform life, business, and the global economy (May)
16. United Nation (2014) World Urbanization ProspectS. 2014 Revision—http://esa.un.org/unpd/wup/highlights/wup2014-highlights.pdf
17. Zarco-Tejada PJ, Hubbard N, Loudjani P (2014) Precision agriculture:an opportunity for Eu farmers-potential support with the cap 2014–2020, June 2014
18. Willis K et al (2012) Printed optics: 3D printing of embedded optical elements for interactive devices. In: Proceedings of the 25th annual ACM symposium on User interface software and technology. ACM, 2012
19. Hierlemann A, Brand O, Hagleitner C, Baltes H (2003) Microfabrication techniques for chemical/biosensors. Proc IEEE 91(6):839–863
20. Hierlemann A, Baltes H (2003) CMOS-based chemical microsensors. Analyst 128(1):15–28
21. Sir Richard Lapthorne, Sir Mark Walport (2013)"Future of manufacturing: a new era of opportunity and challenge for the UK—summary report", 30 October 2013, UK
22. Wind, "ACCELERATING BUSINESS TRANSFORMATION Opportunities and Challenges in the Internet of Things"—http://www.windriver.com/whitepapers/accelerating-business-transformation/Wind%20River-IoT-Business-Transformation-White-Paper.pdf
23. http://www.futuretimeline.net/21stcentury/2050-2059.htm#mars-population-2050s

Chapter 11
Cyber Security: Beyond 2050

Anand R. Prasad and Sivabalan Arumugam

Abstract Technology advancement is happening at a pace that is creating a "technology jet-lag" for the human society. 15 years back, 10 years was considered normal time for research ideas to be productized but today there is barely any gap. Thus in this paper we proclaim that "Cyber 2050" will happen in the (human) year 2020, i.e. 1 cyber year is equivalent to 7 human years. This pace of technological enhancement puts security at stress and thus requiring the industry to re-consider the security model. In this paper, first we present our views of cyber security in the cyber year 2050. Next, we look at the human year 2050 and present our vision of the society and cyber security implications. Basic concepts in this paper were presented during the 100th PhD celebration of Professor Ramjee Prasad (Prasad AR in Cyber security: 2050 and beyond. CITF, Denmark [1]).

11.1 Ready! Get Set!

Beyond Enhancement in technology is happening at a faster pace today than ever before led by ubiquitous mobile broadband connectivity, adoption of connected devices (smartphones, tablets, Internet of things) and in acceptance of new technology by the society. The rapid technology enhancement also brings several security issues with it in the form of cyber-attacks. The economic damage due to cyber-attack leads to short-term as well long-term implications be it financial or otherwise, to the society at large.

A.R. Prasad (✉)
NEC Corporation, Tokyo, Japan
e-mail: anand@bq.jp.nec.com

S. Arumugam
NEC India Private Limited, Chennai, India
e-mail: sivabalan.arumugam@necindia.in

R. Prasad and S. Dixit (eds.), *Wireless World in 2050 and Beyond: A Window into the Future!*, Springer Series in Wireless Technology,
DOI 10.1007/978-3-319-42141-4_11

Telecom infrastructure, a critical infrastructure for any country, is undergoing lots of architectural and service level changes due to increase as well as varied demand for service from customers and service providers. Changes and expansion means increase in footprint for potential attacks and thus cyber security issues. Cyber security is one of the key aspects which need to be considered carefully and it should be addressed systematically in the case of telecom infrastructure.

In this paper, we discuss the cyber security perception and approaches in the past, present and future. In Sect. 11.2, we discuss the Cyber 2050 aspects that focus on future cyber security trends. Evolutionary aspects of Cyber Security and its challenges are presented in Sect. 11.3. Conclusion of this paper is presented in the last section.

11.2 Cyber 2050 in 5 Years

The pace of technology enhancement means that we will see the cyber 2050 happening in 5 years, i.e. by 2020. Let us look at this point in a bit more detail in this section.

11.2.1 Setting the Stage

Mobile communications system has been developing in normal cycle of 10 years for new generation. Look at Third Generation Partnership Project (3GPP) based technologies, Fig. 11.1, with 2G (GSM) in early 1990s going to 3G in early 2000 and 4G in 2010. Now 5G is expected from 2018 onwards. The point here is not the cycle in which new generation is happening but the mobile data-rates—in a matter of less than 20 years the minimum achievable peak data-rate has grown by more than 100,000 times! What does this mean? Connectivity has become ever easier and thus leading to network as a ubiquitous platform to connect and innovate. This platform is the key to technology enhancement—innovation and trial in real-world can be done impromptu thus shortening the time required from idea to market [2].

Up until 3G and its enhancements (Rel. 7) all technology was related to a mobile device connecting to the Radio Access Network (RAN) that in turn is connected to the Core Network (CN). Look at development in LTE era, Fig. 11.2: device to device communication is made possible; aggregation solutions allow a mobile device to connect to multiple base-stations at the same time, provisioning of service is possible without core network. Such development in core technology of mobile communication is a big jump in technological enhancements and a move away from traditional systems.

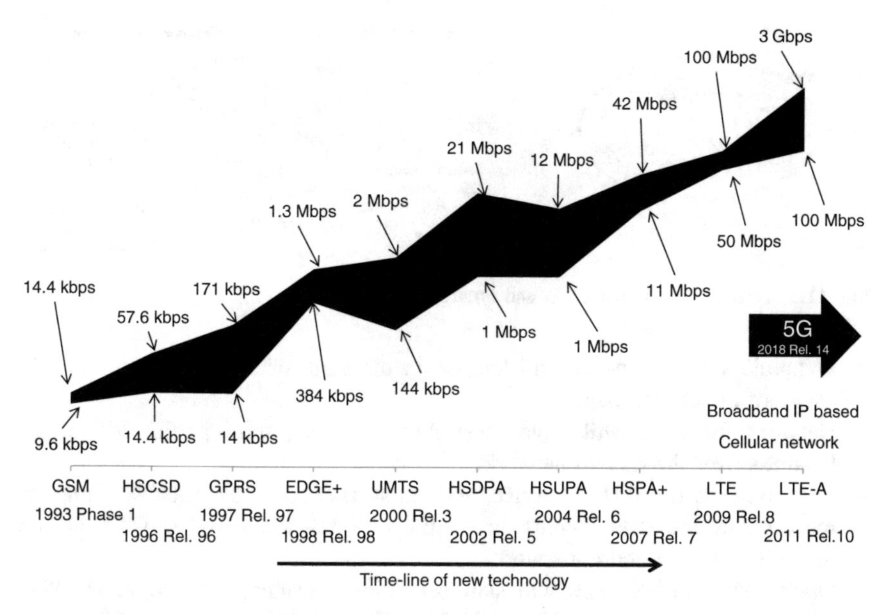

Fig. 11.1 NGMN and Technology introduction [3]

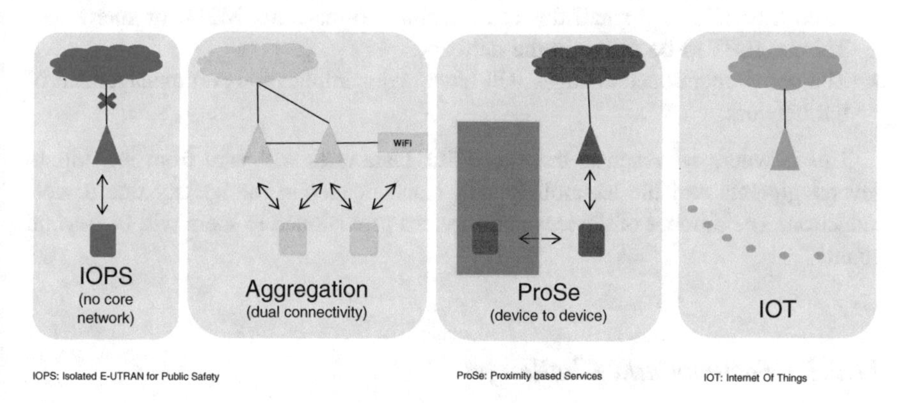

Fig. 11.2 Next generation Core Technology evolution [4]

11.2.2 Evolving Revolution

The next generation that will happen in just a few short years will see tremendous changes, see Fig. 11.3:

- The core network and radio access network will be virtualized (i.e. cloud Ran and Network Function Virtualisation or NFV) which will run on off-the-self hardware connected over the cloud platform.
- Many of the network functions will be based on open source software.

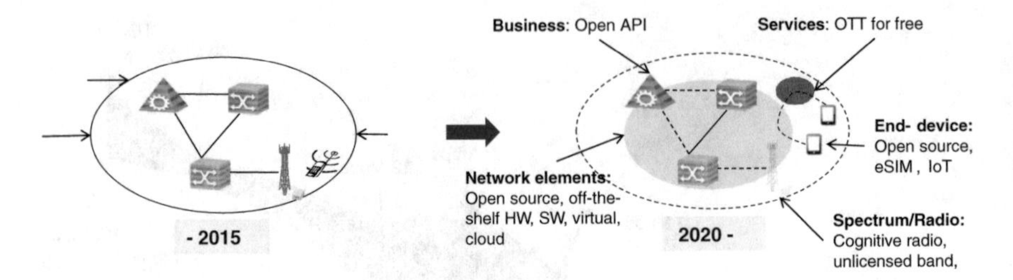

Fig. 11.3 Telecom network: Present and Future

- Dynamic spectrum usage will happen in the form of cognitive radio and the usage of unlicensed band.
- Network functions will have open APIs allowing third parties to develop business over the mobile network.
- Free over the top (OTT) services will exist as before but probably with ever more sponsored data coming in picture else relation between OTT players tightening with mobile operators.
- Open source end-devices will start becoming prevalent with common OSes. Thus a move towards the PC world for mobile devices will be complete.
- eSIM will do its magic in terms of subscription binding with operators but at the same time allowing small devices (machine-to-machine, M2M, or Internet of Things, IOT) to be active in the network.
- The above mentioned changes will have major implications on management of the network.

This is where we reach in the era of 5G. Data rates will vary from few bits to several gigabits and the technology will reach deeper in the society due to cost reduction. The number of devices and services provisioned to them will be beyond count!

11.2.3 Security and Challenges

In case of any traditional IP network, security is provided at the boundary level, by introducing firewall and border gateways. Same concept was applied in telecom network too. But considering the future network architecture evolution current security strategies may not be directly applicable to secure the network and also the services.

The number of end devices and the type of end device aimed to be connected in future network will be in heterogeneous in nature like mobile device, sensors, low processing power and low memory capability tiny sensors etc. Such a low power, low processing capabilities of end devices shall have soft Universal Integrated Circuit Card (UICC) or no UICC. This will create big security challenges to identity, authentication and most of all secure storage of security credentials itself.

Together with this the heterogeneity of technology used will mean that there will always be chances of that weakest security prevail.

Virtualization of the network elements will lead to"no-border" network (see arrows in Fig. 11.3). Connectivity from outside will go to virtual mobile network functions located anywhere in the network in software form. Thus border for security will disappear requiring provisioning of security at each point of the virtualized network much deeper in the network.

Above is only the virtualized part but there is also the cloud aspect. Cloud will allow network functions to move from one location to other in the network. This will lead to issues with storage and maintenance of security credentials of many different levels that is common in mobile networks today.

Security issues will also come from the large number of devices, types of devices, data-rates and service provisioned. Issue related to energy efficiency will become more prevalent and problems of identity as well as identification will need to be resolved. One can also easily assume that newer crypto algorithms will be needed fulfilling range of requirements.

"PC type" open source mobile devices will bring the common IT issues also to mobile networks at a grand-scale.

This all also means that baseline security issues will prevail even more!, Where baseline security means: security and hardening of operating systems (OSes), password management, proper access control, security of logging, TCP/IP stack issues etc. Examples include never changing passwords or passwords that can never be changed (hardwired passwords), update not happening even when issues are identified or solution rolling towards weaker versions.

11.2.4 Looking at Security Solution?

We think that there are two main security solution required, besides those that for sure will be standardized, to cater for at least some of the issues mentioned in previous section:

1. Security assurance: Call it "Network Guardian" covering security audit, security testing (live network, network functions etc.), monitoring with closed loop control, forensics and security operation centre together with early alert systems.
2. Security as a service: The network will adapt to security needs of the service and/or client. Thus personalizing the network for a given need instead of a security solution where one-size-'should'-fit-all. This is also sometimes known as horizontal solution for a given service instead of vertical solution where the network provisions 'the same' to all.

In this period of 5 years we will see that security will not only be a buzz word but it will become normal to cater for it as for voice quality today where any degradation is unacceptable. Besides this we will see new solutions for authentication and key management in mobile communication systems. The variety of services the new system (5G) is expected to provide will mean various level of security provisioning without deprecating the overall system security. Normal implementation issues will exist that will make security assurance ever important.

Privacy issues will and so will there be concerns regarding lawful interception by some. Considerations regarding privacy and security will become common public debate.

11.3 Human Year 2050: The Crystal Ball

Social and economic changes in this world are demanding more and more benefit from technology at large. ICT Technological revolution made it possible that the people are connected always at anytime and anywhere thus leading towards digital/cyber life. In this section let us look at a bit further in future – the human year 2050.

Table 11.1 summarises the technology evolutions of the cyber world. The characters in the table represent the author as a kid in 80s, father in 2015 and with grand-child in 2050.

During early 80s only hundreds of transistors could be bundled and put into a single chip to perform any simple computation. In today's scenario, 2015, advancement in chip technology allows billions of transistor to be put together capable of doing several complex computational tasks within fraction of a micro-second. In 2050 the enhancement will take as to the direction where numbers will not be sufficient to count the transistors else we will be in an era where technology enhancement will make such counting meaningless.

In-terms mobility, during 80s there were only limited mobility was possible for the people as well as in mobile communication technology. Using today's technology, we can achieve communication at very high data-rate while moving at a very high speed with velocities going up to 300 kmph+. The technology is enhanced sufficiently but there is a lot of mobility. In 2050, we envision that even faster physical mobility will be possible but the technology will be advanced

Table 11.1 Technology evolutions: Past, Present and Futures

1980	2015	2050
100s transistors	Billions of transistors	Cannot count
Limited mobility	High mobility	No mobility (needed)
No cyber	Connected	Cyber life
No cyber threat	Coined the term cyber security	Cyber security = Life

sufficiently that we will not be required to move much to be present anywhere. The cyber world will bring virtual reality to all space and depth of human society [5].

During the 80s we barely had connectivity, only a few around the globe were connected although that is when the first cyber-attacks happened. With increase in connectivity we saw security issues growing thus the term cyber security became prevalent since around 2010 leading to, in practical sense, the coining of the term in human society [6]. In 2050s we will see that connectivity will be life itself – the cyber life – but at the same time we are positive that enhancements will happen in development methodologies, and its proper usage, leading to a more secure world.

11.4 Go!

Not only mobile communication technology but the whole Information and Communication Technology (ICT) revolution is providing multiple benefits to the world in-terms of social and economic growth, connecting the world through cyber technology, reducing the physical distance among the people through ICT and much more. However, in security point of view, technology advancement is creating many security and privacy issues, which need to be addressed systematically. We think that the platform ICT is creating will only speed up the pace of technology enhancement. As we move ahead towards human year 2050, we think that security will be embedded in everything. User identification will be simpler using various biometric forms but the tussle between law enforcement (legal and regulatory) and public on grounds of privacy will stay although with better understanding from both sides. Security issues occurring due to implementation will fade away to quite some extent but those existing will be taken care of by monitoring that in turn can create privacy concerns. Thus there is much more to come than we can imagine – this is the start! Ready, steady, get set – Go!

11.5 Conclusion

This chapter highlights some of the key technical evolution expected in cyber year 2050, i.e. in 5 years, and human year 2050. We also summarised the key security challenges which needs to be addressed by us due to the above mentioned technical advancement in the ICT sectors together with thoughts on how we need to face these security challenges. Some of the key points are that technology and technology life-cycle is developing at a fast pace thus leading to security issues that should be catered for. In coming 5 years security will not only remain a buzzword but will be truly considered as the business driver. During the same period 5G will happen bringing several security challenges and enhancements. Looking at 2050, cyber security will be part of life with security being embedded in every system but the tussle between security and privacy as well as law enforcement will remain.

References

1. Prasad AR (2015) Cyber Security: 2050 and Beyond. Professor Ramjee Prasad 100th PhD celebration. CITF, Denmark, June 2015
2. Prasad R (2014) 5G: 2020 and beyond. River Publisher, Sept
3. Prasad AR, Seung-Woo S (2011) Security in next generation mobile networks: SAE/LTE and WiMAX. River Publishers
4. Prasad AR (2015) I will get the star for you "securely". Globecom 2015 industry workshop on 5G security
5. Hitchhikers guide to galaxy: https://en.wikipedia.org/wiki/The_Hitchhiker%27s_Guide_to_the_Galaxy
6. Cyberspace 2025: Today decision and tomorrow terrain http://www.microsoft.com/security/cybersecurity/cyberspace2025/

Chapter 12
Defining the ICT Strategy for Beyond 2050

Sofoklis Kyriazakos

Abstract The sector of telecommunications is one of the flagships of engineering from the time that has been invented, as it was enabling communication that is a major need of humans. In the decades of '90s and early 2000 the rapid development of telecommunications with the parallel evolution of Internet, have significantly influenced people, communities and growth worldwide. During this time, Internet evolution created lots of opportunities, but many of those were not sound and that led us to the dot-com bubble. At the end of the first decade of the 21st century, the Western world experienced a deep recession and so did the sector of telecommunications. On the same time Internet started to recover from the dot-com-bubble period and became the perfect medium for startups. In that period, the evolution of Internet has 'commoditized' telecommunications that served mainly as a medium and not as an innovation vehicle. Nowadays there is a stabilization of dynamics in the ICT world and telecommunications sector is seeking its role and position in contribution to disruptive concepts of the New Digital Era. This positioning and role will be the early definition of its strategy Beyond 2050.

This chapter presents the history of telecommunications, significant milestones and achievements, the New Digital Era, Continuous and Disruptive Innovation concepts and the definition of Beyond 2050 strategy. Throughout the past decades, 'Telecommunications' –as a term- is changing to 'Communications' and is seeking its position in the New Digital Era, an era that is strongly linked with the game-changing evolution of Internet. Continuous and Disruptive Innovation is the holy grail of growth in the New Digital Era and therefore, strategy Beyond 2050 should be focused around such concepts.

S. Kyriazakos (✉)
CTIF, University of Aalborg, Aalborg, Denmark
e-mail: sk@es.aau.dk

R. Prasad and S. Dixit (eds.), *Wireless World in 2050 and Beyond: A Window into the Future!*, Springer Series in Wireless Technology,
DOI 10.1007/978-3-319-42141-4_12

12.1 Recent History of Telecommunications

The word "telecommunication" is compound of the Greek prefix tele- (τηλε-), meaning "distant", and the Latin "communicare", meaning "to share". Obviously, telecommunications is a very old science that served the distance sharing of information with applications that covered needs of all types, including military, governance, business, leisure and many more.

While inventions and novelties have been recorded from the early times, there have been many inventions of significant impact in the recent history of telecommunications, such as the invention of the Telephone by Alexander Graham Bell in 1874 [1]; the invention of the Radio at the end of the 19th century by Marconi [2] and the invention of the Television in the beginning of the 20th century [3]. There are many more inventions but certainly these three are considered to be major, as the enable communication and audio/video broadcasting respectively.

However two sectors of telecommunication have been in the center of focus for the past decades, namely: (a) mobile and wireless communications and (b) wired fixed communications.

12.1.1 Mobile and Wireless Communications

Mobile and wireless communications, are among the most prestigious achievements of the past decades in telecommunications, because of their significant contribution to our daily activities. Mobile communications are characterized in Generations that started to count from 2G and onwards. Today people enjoy 4G services, while in the scientific world there is a lot of research effort and investment in 5G systems.

The Generation of wireless mobile telecommunications includes analog technologies and have been introduced in the 1980s. This Generation was replaced by 2G, which was based on the transmission of digital signals. In the second Generation, the concept of the cellular networks was introduced and the GSM (Global System for Mobile communications) was the most spread standard that has its origins in Europe. In the 2nd Generation, communication was more efficient and new services were introduced, e.g. the SMS service. Initially 2G was based on *circuit-switched* communications, while a few years later the *packed-switched* communications were introduced, i.e. GPRS (General Packet Radio Service) and EDGE (Enhanced Data rates for GSM Evolution) [4]. GPRS and EDGE were the first standards that enabled the penetration of mobile data and Internet in cellular systems. The 3rd Generation (3G) of mobile telecommunications was based on the so-called International Mobile Telecommunications-2000 (IMT-2000) specifications by the International Telecommunication Union (ITU) [5]. Among the differences of 3G compared to 2G, was the evolved radio interfaces that was based on (W)CDMA. 3G enabled a variety of applications, including voice, mobile Internet, fixed wireless Internet access, video calls and mobile TV. In 3.5G additional technologies aimed to increase performance, e.g. HSDPA (High Speed Downlink Packet Access), Evolved HSPA (High Speed Packet Access). The 4th Generation (4G) includes technologies like HSPA+21/42, the WiMAX (which is now obsolete), and LTE (Long Term Evolution) and achieves much higher data speed compared to 3G [6].

12.1.2 *Wired Fixed Communications*

Wired communication are end-to-end fixed communications and exist for a much longer period compared to mobile and wireless communications. The evolution is also significant and that can be easily observed in the type and quality of the services currently offered, compared to the past decades. The big breakthrough happened when fixed telecommunications were upgraded into digital up to the last mile. Backbone systems were evolving and new technologies boosted core networks, to provide infrastructures for large volume of data, including demanding applications in terms of time delivery.

Among these networks, it is worth to briefly describe topologies and infrastructures that facilitate Internet network. Internet is a global set of networks that interconnect in many topologies to result in the "Internet". The backbone networks of Internet do not rely on any central control and this is among its major characteristics and advantages. The Internet backbone is based on several networks of various owners with redundancy that rely mainly on fiber optics that allow fast data speeds and large bandwidth.

Internet usage has been significantly increased in the past decades and has surpassed 3 billion citizens and around 50 % of world population (Fig. 12.1).

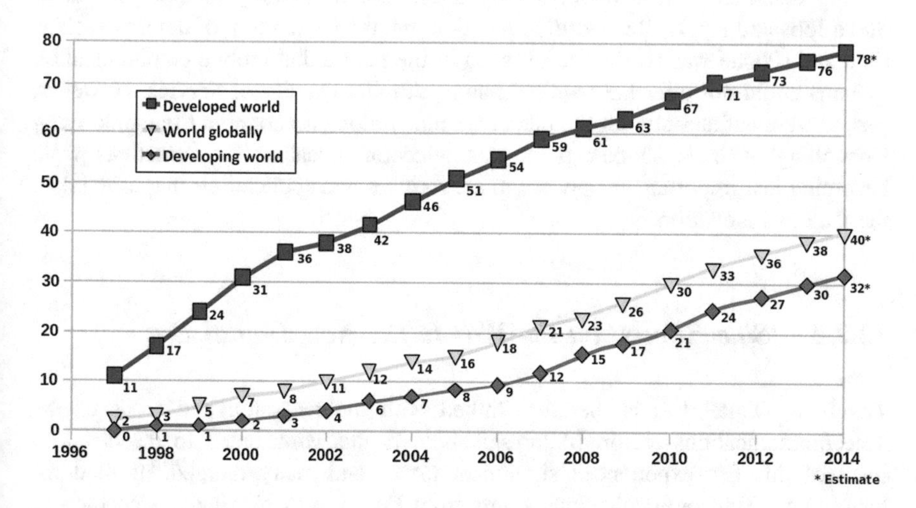

Fig. 12.1 Internet penetration [7]

12.2 Current Positioning and New Digital Era

In the previous sections the recent history of telecommunications was discussed, as well as the recent achievements of wired- and wireless-communications. The evolution was significant and that can be easily understood if compared with other technologies and sciences, e.g. automotive-, space-, pharmaceutical-industry. This evolution has been a driver of the telecommunications industry that was a significant percentage of the GDP of the Western world. The rapid development though is naturally followed by a recession, which is the case in the telecommunications sector too. The recession of the telecommunications sector is observed in the time of the New Digital Era.

12.2.1 Defining the New Digital Era

There are many definitions of the New Digital Era. Based on the one that describes the shift of focus among ICT sectors, *the New Digital Era is the time after telecommunication technologies and infrastructures reached a certain level of maturity and considered to be commodity in the perception of the users, while services and applications became more important*. This is a paradigm shift of telecommunications that is now considered to be an enabler of services and the underlying technology that is mandatory for the economy growth.

The perception of telecommunications being a *commodity infrastructure* is obvious, given the fact that new radio interfaces and protocols have less contribution to the user experience, compared to the contribution of innovative services and applications.

Any discussion on technology innovation should certainly include the era of Steve Jobs and Apple. It is worth to mention that the penetration of devices such as iPod and iPhone was so significant, even in the period that mobile communication systems could not offer the required bandwidth and Quality of Service. However, people were enthusiastic for the perceived innovation and not about the underlying technology. This is another proof that telecommunication infrastructures were becoming less important as opposed to the services and applications that have taken the lead of innovation.

12.2.2 'Side-Effects' on the Way to the New Digital Era

The New Digital Era is therefore linked with the recession and crisis of the Telecommunications sector. Major stakeholders that contributed to the development of this Era experienced significant losses and many merged, or filed for bankruptcy. Telecommunications giants from Europe and North America such as

NOKIA, ERICSSON, SIEMENS, NORTEL, ALCATEL, LUCENT and many others were the victims of this evolution. That was somehow a natural result, as the attention was not anymore on infrastructures but on services and applications. A few of the mobile phone and infrastructure vendors tried to pivot and address effectively the area of applications but in most of the cases that was not successful. In the time of recession of the European and North American telecommunications industry, Chinese companies, such as Huawei and ZTE achieved a high growth, mainly due to their capacity and competitiveness.

Furthermore a number of applications gained the attention by the power of crowdsourcing and the smart use of Internet and telecommunication infrastructures. That had as a result the creation of the largest taxi company that has not even a single driver, namely UBER; the creation of a big accommodation service that has not even a single hotel, namely AirBnB and many more examples. This can be considered as a 'side-effect' of the New Digital Era too, if the impact on traditional incumbent players is considered.

12.2.3 The Bright Side of the New Digital Era

In the beginning of the New Digital Era, Internet experienced a huge development and so did the whole industry that relied on it. The New Digital Era includes smartphones, game consoles, smart TVs, wearable devices and of course Web2.0. An important aspect that was considered by the R&D industry was the convergence of Telecommunications with Information Technology, shortly ICT (Information Communication Technologies). Researchers of the two sectors came closer and created value for the end users. In this value chain there was room for more contributors than engineers and scientists, for example innovators, marketers, designers and people from the financing sector that intervened in the production process and created tangible outcomes for the end users.

Meanwhile, other business sectors have been hit by the Internet, such as above-the-line advertising, hard-copy journals and books and of course the music industry. Business models have totally changed and most companies were not able to pivot and avoid the catastrophe. In the music industry, in the 80s people were relying on LPs and CDs, then this changed into downloadable pieces of music (e.g. iTunes), which is now under a new change process to subscription-based music (e.g. Spotify).

The evolution of music business model is shown graphically in Fig. 12.2.

One can see that in the past century there have been decades that were characterized by the use of specific technologies and media, e.g. vinyl, cassette, CD, while in the last few decades one observes the appearance of services that introduce different business models. These business models have either been achieved through new technologies, or even through disruptive models, e.g. Napster.

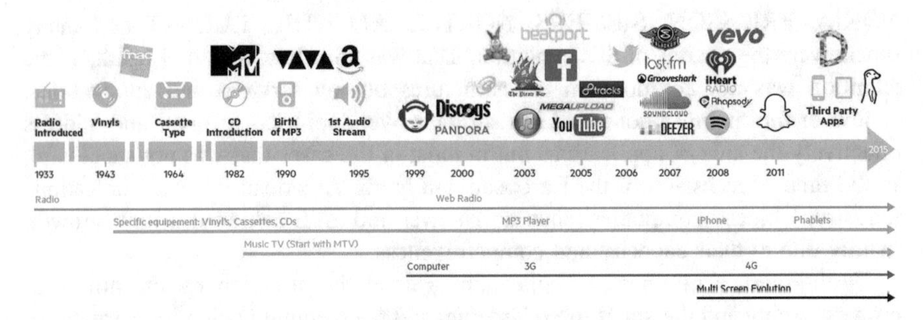

Fig. 12.2 The evolution of music business model. *Source* http://www.themediashaker.com

12.2.4 New Digital Era Ecosystem and Disruptive Innovation

The New Digital Era that people are currently experiencing, provides an ecosystem for people to create value and prosper. This ecosystem requires continuous innovation management and pivoting based on actionable metrics. Unlike past decades, individuals can easily setup business with the help of the available ICT infrastructures and identify niche markets that have needs. Access to information through Internet and KPIs and metrics that measure the effectiveness of any application, allow the pivoting to optimize a concept and transform it into a winning application. An evangelist of this strategy is Eric Ries who describes how to build, monitor and optimize an online concept in the Lean Startup [8].

Disruptive innovation does not require necessarily knowledge of the telecommunications infrastructure and most of the times, entrepreneurs take the underlying infrastructures for granted. Disruptiveness requires open-minded approach in identifying or even creating market needs. There are many paradigms of young entrepreneurs who started a garage-business and in a few years' time they *conquered* the world. Among them is obviously Mark Zuckerberg of Facebook that has penetrated to almost every data-connected place on earth.

Innovation alone might not even be enough though, to produce profitable and sustainable business. The question is how disruptive a new concept is. Nowadays, people not only seek ways to enter and break barriers in existing market, but to create their own markets too. This is well described in the Blue Ocean Strategy, a well-known business approach, from the Institute of INSEAD [9] that is available from Harvard Business School Press as a book [10]. Based on the Blue Ocean Strategy many companies managed to create new markets and make competition irrelevant in their attempt to become sustainable and profitable.

In a similar way, Apple managed to become a game-changer by introducing iPhone and by creating a market based on a disruptive business model that involved crowdsourcing mechanisms to develop, sell and profit from applications. Crowdsourcing is proven to be a strong tool to engage people in co-producing content in a win-win relation with the service provider. Foursquare [11], for example is providing personalized location-related information for subscribers of its service. The subscribers feel the *social obligation* to check-in in places they visit and create content. This crowdsourcing process results in a rich database of information for the Foursquare users, while the service provider benefits from disruptive revenue models on top of this data. Crowdsourcing has a significant power and growth and is expected to contribute to the democratization of communications in the near future.

Disruptive innovation, such as the above mentioned examples, requires to take a step back and begin with the end in mind, as Stephen Covey suggests in [12]. It is the step that any pioneer should make in order to see where he/she wants to go aiming to create valuable services or products and new markets. Disruptive innovation is linked with the hype of start-ups that lead young people to innovate and create their businesses. This is not an easy process, as innovation itself is not the only element for the success of a new venture. In any case this is an evolution in the traditional entrepreneurship that was excluding those in the past who were not privileged.

12.3 The Next Big Thing and the Role of Telecommunications

Disruptiveness is a key element in innovation creation and management. This is the opportunity of telecommunications to take again the lead and from commoditized state and business enabler, to become the means to differentiate from the competition. M2M communications, Internet of Things or even the Internet of Everything concept of Cisco, are examples of how telecommunications is coming back to a leading role. In 5G, there are many concepts under discussion, among them the cooperation of networks and the availability of multiple operators that can even be subscribers. Such disruptiveness, having its fundamentals in the new 5G concepts, can change the rules of the game.

The next big thing is not necessarily a new radio interface, a smart-phone, or an application. In any case, the next big thing will be a game-changing concept that may follow a user-centric approach, where the user will have his/her share in the value that he/she adds. This will create a new era, where users will not be appreciated only as customers, but they will be rewarded accordingly. The new role of users will empower them; thus creating a new market and business model. This new business model is the holy grail of ICT entrepreneurship for individuals and enterprises.

The next Big Thing can exploit the steps towards 5G and the ICT infrastructural challenges of Future Internet that set the threshold very high to support larger traffic volumes of 3 orders of magnitude within a few years' time. This is a unique opportunity to come up with highly competitive concepts that can be the ground to support the traffic evolution, minimizing the time-to-market period for R&D outcome. User-driven concepts that satisfy demanding needs, such as comparable wireless and fixed access rates, managing the increasing energy consumption requirements and the performance characteristics introduced by multimedia applications can be linked with the need to have reduced CAPEX and OPEX of networks, as a natural consequence of the technology advances. During the period of ICT convergence, end-users may have multiple roles, i.e. consumer, creator and even operator of services and therefore new win-win business models can be created to guarantee the sustainability of the large investments in the new era of Internet.

5G could be the enabler of the next *Big Thing* based on intelligent infrastructure networks. The minimum requirements for the intelligent infrastructures balance between performance and convergence aspects to support heterogeneity, such as scalability, interoperability, reliability, availability and adaptability. The 5 major challenges identified by EC through R&D activities are: (a) very high data rates, (b) very dense crowd of users, (c) very low energy cost and massive number of devices, (d) mobility, and (e) very low latency. From these challenges, following scenarios are derived: (a) Amazing fast, (b) Great service in a crowd, (c) Ubiquitous things communicating, (d) Best experience follows you and (e) Super real-time and reliable connections, [13].

Based on these challenges several scenarios can be defined; for example, the case of Ubiquitous Smart City using ICT to make its infrastructure, public services and components, efficient, interactive and more aware citizens. In this Ubiquitous Smart City, very fast broadband could be reaching homes and this has to be extended by operators to the city area, regardless of the mobility status of the users, thus embracing a wide variety of OTT applications to both commercial users and the general public that are looking for services beyond triple-play. The key challenge is how to create game-changing opportunities and transform usual ICT business cases to societal value cases maintaining the self-sustainability and self-management of the infrastructures.

12.4 Strategy Definition for Beyond 2050 and Conclusions

When it comes to the ICT strategy definition for Beyond 2050, the natural evolution of the New Digital Era on the users, infrastructures, business sectors and business models has to be assessed first. The users are expected to increase significantly both in terms of age groups and geographical distribution. It is expected that children at the age of 5 will be involved in STEAM (Science, Technology, Engineering, Arts and Mathematics) education programs and this will be achieved partially by ICT devices and infrastructures. Moreover, senior citizens in the age group of 74–85 or higher, will be the people that are today in the age group 25–34, therefore the deep involvement with the technology is inevitable; not to mention the various digital services that will be available and personalized for these ages. As long as the geographical distribution is concerned, it is expected to cover all continents and reach penetrations than 80 % worldwide. Today, the estimated Internet penetration in Europe and Oceania is between 70 and 75 %, in North America around 88 %, in Middle East and South America between 50 and 60 %, in Asia around 40 % and in Africa less than 30 %. The overall Internet penetration is a bit higher than 45 %. The areas where Internet penetration is low and population high is Asia and Africa. In both continents there are significant project for Internet infrastructures and initiatives for free basic internet [14], as well as education programs. Overall it is expected that in 2050 people will be living in a digital world, where Internet will be pre-requisite for any business sector. The increase of Internet penetration in the next 35 years is expected to connect more than 3 billion users! This number is enormous and so are the needs the opportunities.

The infrastructures are also expected to grow dramatically. First of all, they should be able to accommodate traffic from another 3B users that will be much more demanding, compared to the past. As far as the dimensioning of infrastructures is concerned, people should not only think of today's applications. A few decades back, people thought that kbps would be enough to entertain people and serve their needs. Today individuals have bandwidth requirements of several Mbps and enterprise requirements are in the order of Gbps. Certainly the bandwidth is a key factor, but mobility is of major importance too. Mobile communications of 4G and beyond offer high data rates on the fast move, but still they are not mature enough to guarantee certain QoS. In the years to follow, Gbps bandwidth is envisioned in almost every rural area of the planet, with guaranteed QoS that will allow mobility. Internet will be available and affordable in any vehicle, including planes, boats and underway trains. Every device, active or passive, will be connected and accessible over the Internet in a secure manner. The IoT era will offer a great number of applications that will be smoothly integrated in the everyday life.

Furthermore, there will be no business sector that will not be significantly influenced by digital communication technologies and Internet. Even today, sectors that were not traditionally linked with advanced technologies, are now connected (Fig. 12.3).

Business models are also expected to become simpler, by including more stakeholders and by distributing revenues based on meritocracy. It is expected that similarly to the competition experienced in telecommunications when the monopoly stopped, the competition in the introduction of new business models will rise in a similar way by rewarding the engaged and loyal users. Advertising was a major revenue stream for many operators and media so far and is expected to remain at a high level. Nevertheless, the income will not be distributed only among operators. The New Digital Era enables a detailed measuring of the impact of each type of advertisement and therefore anyone who can contribute in branding and sales of a merchant or vendor can be rewarded accordingly. The advertiser can therefore become the end-user, who could even be at the same time the content- or app-provider, as it happens for example with smartphone users who are application developers, exploiting open or closed marketplaces addressing a large community of end users.

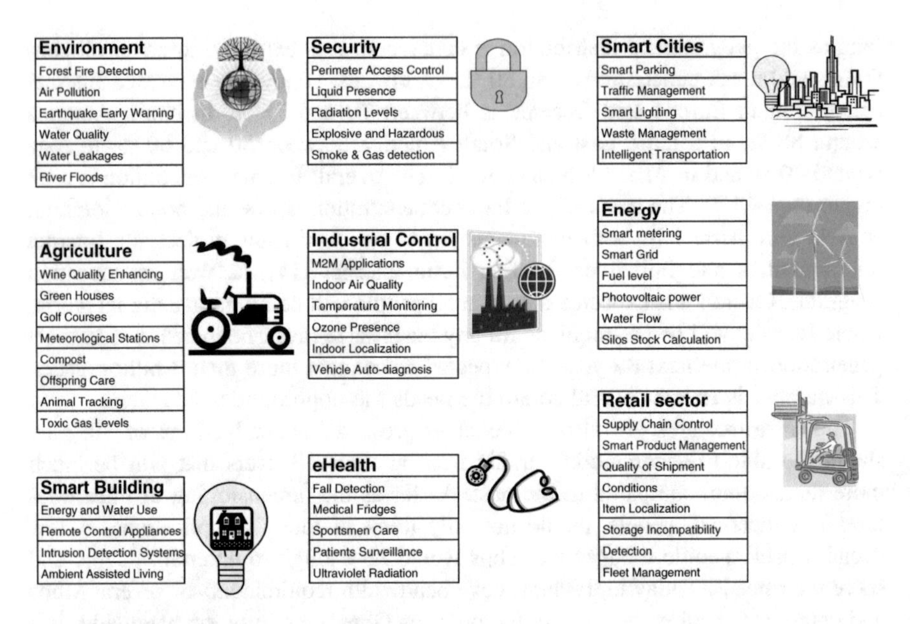

Fig. 12.3 IoT applications in various business sectors [15]

Another important parameter for the definition of the Beyond 2050 strategy is the wealth of Big Data. The amount of data that operators are collecting is growing rapidly, given the ability to easily track events and information, reaching the level of Petabytes. The challenge of Big Data will therefore be to exploit and mine data information to result in *actionable data*. For example, the location of all vessels around the world can produce data of similar volume, but the combination with the type of the fleet and load can provide important metrics for macro-economical values, such as exports of goods from China related to the GDP of the country.

On the downside of the evolution of the Digital Era a parameter that will be important for a Beyond 2050 strategy definition will be the *data fatigue*, as an outcome of the volume of information that are made available and consumed by the people. That increases the stress and might have negative impact too. Another side-effect is the decrease of the time for socializing in the real world and not online through social media channels. That may influence the value system of the society and generate new trends that will emphasize the need of maximizing the time spent with family and friends, detached from communication devices and media.

In this changing world, where communication infrastructures will be available everywhere, for everyone, with unlimited resources; in a world where narrow-minded corporate policies will be substituted by disruptive business models, it is a clear need to define the strategy for Beyond 2050. Many believe that Beyond 2050 is far away to define a strategy; this should be considered however as an opportunity to set the fundamentals today. These fundamentals are summarized into the following 10 strategic steps:

1. Design applications and services assuming infinite networking and communication resources;
2. Follow a user-centric approach by engaging the user in an agile development manner;
3. Exploit the power of crowd-sourcing by incentivizing contributors;
4. Embrace open-source in software and hardware;
5. Develop services and applications thinking out-of-the-box;
6. End-to-end integration following an anything-to-anything approach;
7. Rewarding mechanisms for all users based on meritocracy;
8. Beat the fear of openness in systems and ideas;
9. Democratize communications and commoditize networks and systems;
10. Create a dynamic ecosystem that follow the lean startup paradigm.

These rules should apply to any ecosystem involved in Beyond 2050 technological matters and that will be the driver for the economic growth. Communication infrastructures, technologies and services should become the dominant part of the world's GDP, creating a promising environment for the citizens to prosper.

References

1. H.Res.269—Expressing the sense of the House of Representatives to honor the life and achievements of 19th Century Italian-American inventor Antonio Meucci, and his work in the invention of the telephone. 107th Congress (2001–2002). U.S. House of Representatives. Retrieved 7 Feb 2014
2. Marconi and the History of Radio (2004) IEEE Antennas and propagation magazine 46(2):130
3. Abramson A (2003) The history of television, 1942 to 2000. McFarland, Jefferson, NC, and London. ISBN 0-7864-1220-8
4. 3rd Generation Partnership Project (2015) 3GGP TS45.001: Technical Specification Group GSM/EDGE Radio Access Network; Mobile Station (MS)—Base Station System (BSS) interface; Radio Link Control/Medium Access Control (RLC/MAC) protocol; section 10.0a.1—GPRS RLC/MAC block for data transfer. 12.5.0. Retrieved 2015-12-05
5. International Telecommunication Union (ITU) (2013) IMT-2000 Project—ITU (April)
6. LTE—An End-to-End Description of Network Architecture and Elements (2009) 3GPP LTE Encyclopedia
7. International Telecommunication Union (ITU) (2015) Individuals using the internet 2005 to 2014 (May)
8. Ries E (2011) The lean startup. ISBN 978-0-307-88789-4
9. http://www.insead.edu/blueoceanstrategyinstitute/home/index.cfm
10. Chan Kim W, Mauborgne R (2005) Blue ocean strategy. ISBN 1-59139-619-0
11. https://www.foursquare.com
12. Stephan Covey (1989) The 7 habits of highly effective people. ISBN 0-7432-6951-9 2013
13. https://www.metis2020.com/
14. https://info.internet.org
15. Prasad R, Poulkov V (2015) Resource management in future internet. River Publishers

Chapter 13
Multi Business Model Innovations Towards 2050 and Beyond

Peter Lindgren

Abstract The development and innovation of business models to a world of 5G—is indeed still a complex venture and has not been widely researched yet. Numerous types of technologies are expected and in these years proposed and in the "slip stream" of these the persuasive technologies and persuasive business models are expected to gain more and more importance in a world of a 5G. The application of the persuasive BMs will begin with 5G roll-out in the next 5–10 years and is expected to be common place by 2050. The development of persuasive technologies gives hopes to realize persuasive business models where business models can act and be innovated secure. Although persuasive business model use persuasive technologies and are basically created to "persuade" for a certain behavior in accordance with the strategy of the BM and business they should be constructed secure. However there are still some steps to take before we reach this stage and a deeper understanding of how businesses really can use persuasive business models and what they really can do with persuasive business models. The chapter gives a conceptual futuristic outlook secure persuasive business models on behalf of inputs from SW2010—SW2015, lab experiments in the MBIT and Stanford Peace Innovation Lab together with state of the art persuasive business model and technology research. The chapter touch upon what we can expect of persuasive business Models and persuasive business model innovation in a future world of 5G.

"In the past ten years the number of sensor-, wireless and persuasive technologies in our everyday life, have increased many-fold. We are now moving fast towards a world of 5G which obviously will by standard have embedded persuasive technologies [1–3]—and therefore it will soon be a reality to business to deal with these technologies. "The study of these persuasive technologies, and how they affect our lives and routines are still very young—and we know little about how they will

---

P. Lindgren (✉)
Aarhus University, Aarhus, Denmark
e-mail: peterli@btech.au.dk

R. Prasad and S. Dixit (eds.), *Wireless World in 2050 and Beyond: A Window into the Future!*, Springer Series in Wireless Technology,
DOI 10.1007/978-3-319-42141-4_13

affect our future lives—but we know that what we can expect of persuasive technologies cannot even we imagine today [3, 4].

Researchers, business and public players alike are keenly devoting themselves to understanding how these different persuasive technologies might be designed, so that desirable technologies, behaviour and not least "business models" are obtained and can be created, captured, delivered, received and consumed–hopefully secure and sustainable—in a new 5G world. There is large expectation to that 5G and persuasive business models will enable better business and also global economy. Especially healthcare and well care sector expects much of persuasive technologies to overcome some of their big economic burden due to amongst others a growing elderly population and increasing medicine costs.

The power and importance of persuasive technologies embedded in a multitude of business models is therefore obvious!—and to some extent would some say—on the dark site scary—if not secured and lead in a valuable and sustainable directions. However the evolvement of persuasive technologies and persuasive business models are not to be stopped and their impact will be enormous in the future and even be clearer very soon!

13.1 Persuasive Business Models and Business Model Language

Today most businesses are not really able to download, see, sense, act-do, scale and globalize on their Multi Business Model Businesses and Business Model Ecosystems. Many businesses are what could be classified as "Business Model analphabetic" and are not really understanding and using their full BM potential. This results in a large waste of potential, resources and competences.

Alignment and an accepted business model language are one reason to this "loss"—preventing business to take the next step in business modelling. A common agreed business model language is therefore highly needed to make it possible for business to communicate their Business Model dimensions and components with one another.

In a world of 5G this language will be even more important. If all could agree upon a common business model language then business model innovation and the creation and use of persuasive business models could really take off.

Today most businesses are just able to see a small part of their BM value exchange and often just from one viewpoint in a 2D mapping as we illustrate from one of our research projects in Fig. 13.1.

Business can with big efforts today download and to some extent see values exchanged (green balls) but mostly often just on adhoc basis and just for one business model inside their business at a time. If they had agreed on a BM language internal the business this could be different.

5G will enable us with data and abilities to "see" the complex world of Business Model Innovation. Due to all the interactions that 5G technology will give us

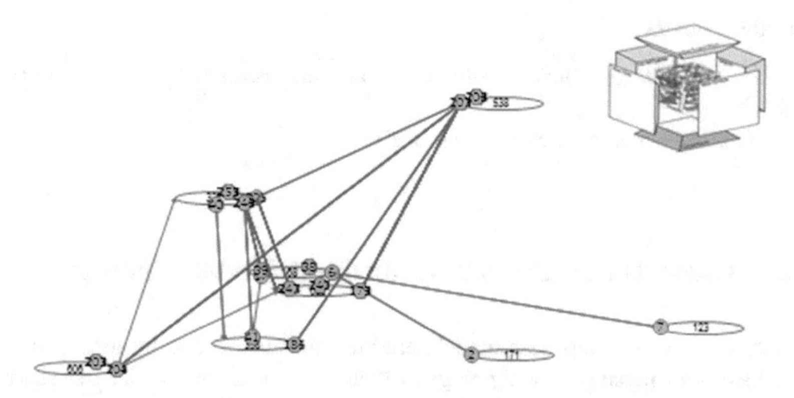

Fig. 13.1 2D Mapping of one BM value exchange sequence from an industrial business

combined with advanced visualisation technology we will be able to "see". It will provide us with enormous amount of new data, knowledge and insights, which we will be able to understand when we have agreed on "a common language" and where to look.

Having taken the first step into "unwrapping" this new knowledge in small lab experiments with a small part of a business with just 15 different BM's the picture shows us a rather complex and less operational picture of the value exchange between these business models (Fig. 13.2).

Understanding this complex Business Model value creation, capturing, delivering, receiving and consumption process, of both tangible (full line) and intangible (dotted line) value exchanges between Business Models will however be essential to create influencing BM—"to create persuasive BM".

Also from a security perspective, business and society will face tremendous challenges in understanding and leading persuasive business models and leading the future business model innovation processes to become sustainable.

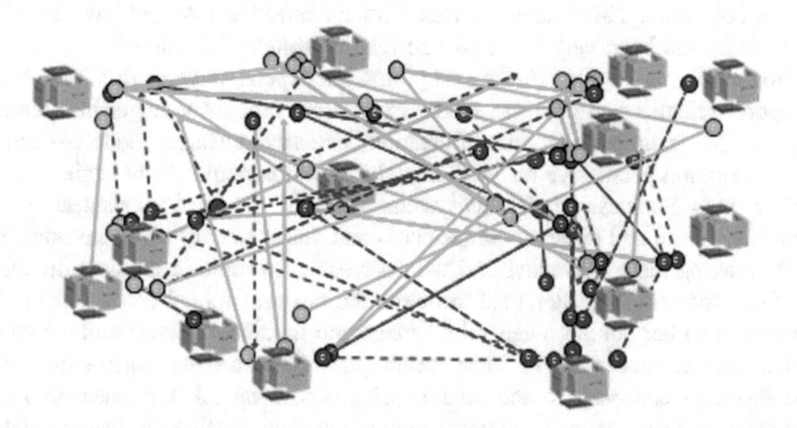

Fig. 13.2 2D Mapping of a multitude of BM value exchange sequence between BM's inside a business

In this context

How can we achieve security, profitability and other values of persuasive business models?
How to embed security in persuasive BM's?

13.2 Towards the 'Persuasive Business Model' Concept

By nature persuasive business models are built persuasive and by nature they are embedded with persuasive technology, which if not lead can hinder the vision of new and better business models in a sustainable world of 5G to become true. Our hypothesis is that persuasive business models will be in common use by the year 2050. By this time, nobody knows which mobile wireless standard(s) will be in use, however definitely beyond 5G. Therefore, we expect that the real application of the persuasive BMs would begin with 5G roll-out in the next 5–10 years and will be common place by 2050.

Both users, customers, networkpartners, employees and businesses of today—in general will find initially those persuasive business models and persuasive business model innovation processes highly risky, foreign and radical related to existing BM's and BM innovation processes–that they've been experimenting with in their past.

The history of clarifying the persuasive business model concept is however relatively young. As 3G and 4G based business ecosystems emerged, many business (Google, Facebook, Amazon, Ebay, Zinga, Blizzard began rethinking their business model and business model structure [5]. They began to build in persuasive components (motivating colors, text, tabs, sounds), dimensions (value propositions, value chain activities, relations and networks) even BM's which could "motivate"—some would say "persuade"—users, customers, network and employees to certain behaviors. These attempts were initially rather simple and have up to for some years ago been very harmless and relative simple constructed.

However when highly professional promotion experts, phycologists, sociologist, computer scientist and business model experts are brought together in interdisciplinary BMI teams with the aim of creating persuasive business models—then the next generation persuasive business model innovation could really begin and find their way to Business Model Ecosystems (BMES). Persuasive business models demands several and different competences and this is exactly where investment in BMI is taking place nowadays and will increase even more in the next decade.

Many authors have attempted to define the business model concept [6] but as mentioned earlier not yet a general language and framework concept of a business model has been accepted [7]. Most authors have taken a rather narrow innovative and financial approach to the context of business model and business model innovation—a very general and transaction oriented approach to business models. It seems that most (if not all) authors have not done the combination of the terms:

'business model' and 'persuasive' technology seen in an interactive and relational perspective.

Previously

A business 'business model' was defined *as a building platform that represents the business strategic, operational and physical manifestation.*

We claim that the real challenge for business model "designers" is to first identify all dimensions of the business BM's—download, see and sense them and their key dimensions, components and key relations to their business both for 'AS-IS' and 'TO-BE' business models before they can even begin to understand, innovate and make them interactive and persuasive [8–10].

The persuasive business model approach bring the business model into a more advanced step compared to previous development—a new era—by answering the question:

What if we really could use the sensors, data and knowledge we gather and could gather knowledge in the interaction with people and things—to really "influence" or "persuade" for certain behaviors of people and things?

It has been argued that a business model framework must be reasonably simple, logical, measurable, comprehensive, operational and meaningful. The persuasive business model we propose have to meet the same approach. We proposed the business model previously to be constructed of 7 dimensions (Table 13.1) to represent all generic dimensions of any business model [11]. The persuasive Business Model we propose also as related to these 7 dimensions.

Our previous research [12] show that a persuasive business model at an optimum adapt a multi business model approach combining and relating different "ingredients" from more than one business model. It means that persuasive business models

Table 13.1 Generic dimensions and questions to any persuasive business model

Dimensions in a generic BM	Core questions related to a generic persuasive BM
Value proposition/s (products, services and processes) that the business offers (Physical, Digital, Virtual)	What are our value propositions?
Customer/s and Users (Target users, customers, market segments that the business serves—geographies, physical, digital, virtual)	Who do we serve?
Value chain [internal] configuration (physical, digital, virtual)	What value chain functions do we provide?
Competences (assets, processes and activities) that translate business' inputs into value for customers and/or users (outputs) (Physical, digital, Virtual)	What are our competences?
Network—Network and Network partners (strategic partners, suppliers and others (Physical, digital, virtual)	What are our networks?
Relations(s) e.g. physical, digital and virtual relations, personal (Physical, digital, virtual)	What are our relations?
Value formula (Profit formulae and other value formulae (physical, digital, virtual)	What are our value formulae?

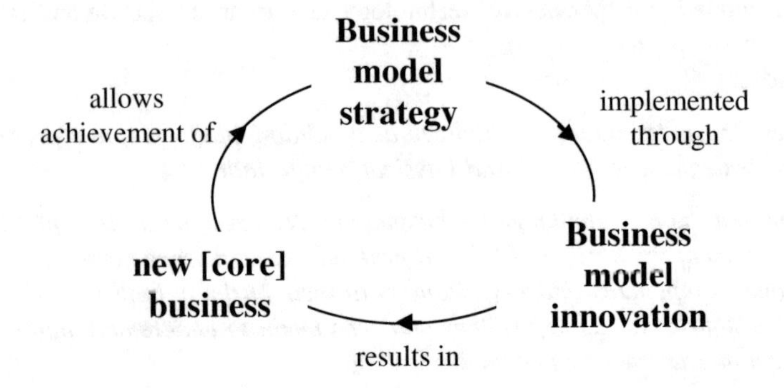

Fig. 13.3 Core business, business model innovation and business strategy—Taran et al. 2009

are in relations with other business models and most of them have some kind of embedded persuasive technology built in.

Although some researchers [1, 13] include strategy in their business model definition, it was conceptually found clearer in our previous BM research and approach [12], however, also in view of the role of business model innovation process, to at least analytically distinguish between strategy, dimensions, components and core BMI processes of the business model [14] as shown in the model beneath (Fig. 13.3).

Relating—to the previous discussion—it can be argued that strategy [2, 15], are embedded with the persuasive business model approach, providing the larger platform for the business who strategically have been decided to be based upon and act with one or more persuasive business models.

With other words—a persuasive business model includes an interactive strategy vision, mission and goal(s) where the business model seeks to achieve with the impact of changing the business models dimensions to change the user, customer, network, employee or things—even all.

Taran et al. [14] found 3 approaches to 'define' when a business model innovation is a radical change. In the way a business do business [16]. The first approach regards any change in any of the [core] business model dimensions or the relationships between them as a form of business model innovation. The second approach, in line with Abell [17] and Skarzynski [18], involves considering the number of dimensions in the BM that are changed. The third approach defines innovativeness in terms of, what might be called, the reach of the business model innovation. A suitable scale to measure the "new to whom" of a business model innovation could be one ranging from new to the business, via new to the BMES segment and new to the BMES and finally new to the world.

A three-dimensional space then emerges, which helps in qualifying the innovativeness of a business model (Fig. 13.4):

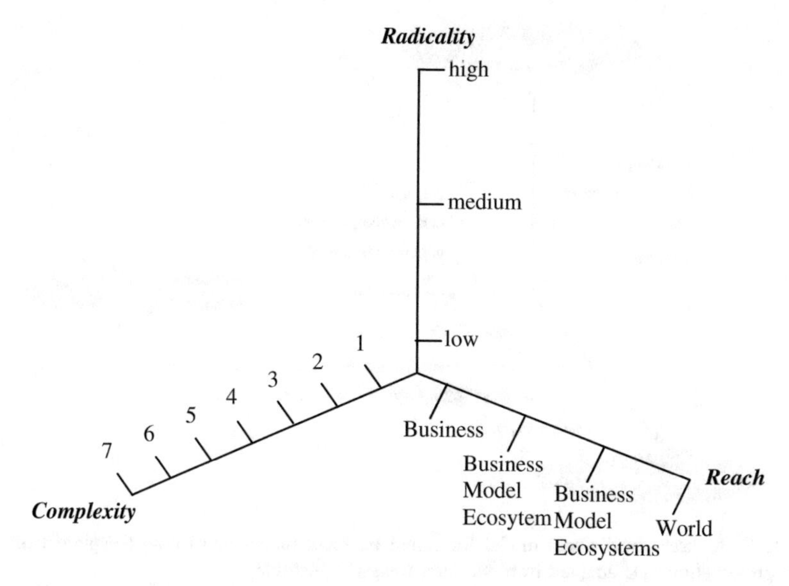

Fig. 13.4 A three-dimensional business model innovation scale inspired by Taran et al. [16]

- **Radicality**—(how new?) incremental versus radical of each BM dimensions?
- **Reach**—to whom the business model innovation is new?
- **Complexity**—how many dimensions of a BM are changed simultaneously?

Because of the approach of how to develop 'a persuasive business model environment' the persuasive business models of business have to be structured and lead in a much more strategic way than classical BM's—which eventually means that the business have to commit itself to a persuasive BM strategy approach…

- that a persuasive business model by nature have strategy built into the business model.
- the persuasive business model are strategically designed with the aim to change users, customers, networkpartners and employees behavior via its value proposition(s) acting interactively together with the other 6 BM dimensions and related BM's.
- the persuasive BM is interactive by nature and at best functions optimal in business and together with business models that are prepared for this.

13.2.1 A Proposal to a Secure Persuasive Business Model in a World of 5G

Lindgren et al. [10] back in 2010 introduced a first proposal for how the environment for business model innovation could look like in a future mobile and wireless communication technology society (Fig. 13.5).

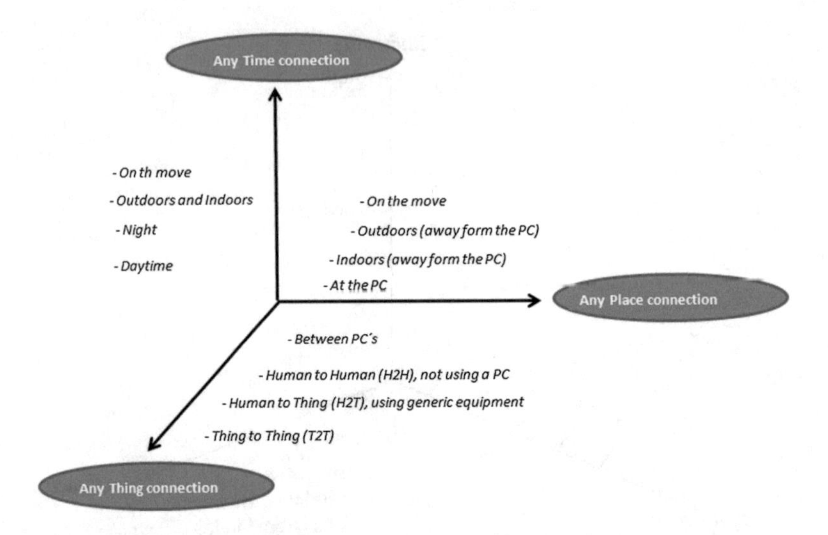

Fig. 13.5 A three-dimensional model for future business model innovation (Inspired from TU Delft presentation ITU adapted from Nomura Research Institute)

The persuasive business model will in this context by nature try to attach to anything, anybody, anywhere and anytime—with the overall aim to "persuade". 5G is expected to be the backbone of most persuasive business models and persuasive business model ecosystems.

First, businesses will use the expected advanced 5G sensor network to link into all 7 dimensions of any business models. Second, the majority of persuasive business models will be and act as network of persuasive business models. Third persuasive BMI will be carried out via advanced persuasive technology, which means that businesses become increasingly dependent on 5G network to extend their persuasive business model innovation and development to the BMES. Fourth anything, anybody, anytime in any place will be linked persuasively together.

The persuasive business models based on networks of different persuasive business model technology platforms, software and BM ecosystems we expect will be a reality soon—but business models will not only act physical and digital—but much more virtually—in a full integration between the persuasive physical, digital and virtual business models. 5G—especially the advancement of sensors and sensor technologies—enables indeed some new dimension on persuasive BM and persuasive BMI, which raise the question of—How to enable security and persuasive business models at the same time?

5G enables both the vision of creating persuasive business models any time, any place, with anybody and anything but also stress the question—how to be secure any time, any place, any body and anything when everything and everybody are persuasive. Today human beings and human bonds have in most case actively "forced" security and a security profile—if it has been possible to do so. Numerous systems of security have been developed—but as presented on the SW2015 [19] today nobody can be 100 % secure anytime, anywhere, any place—with anything and anybody. How much security can 5G eventually give us and do we wish actually to have security anytime?

In the future all human beings, all things at any time and in any place will have the possibility to act persuasively and to some extent at the same time secure themselves automatically. However the network based persuasive business model of tomorrow will consist of business models with all kinds and many types of strategies, security structures and offers—including virtual, digital and physical value propositions and security systems. Who and which should we trust? In this context we—as pervasive business model provider and developer we have to face a world where we innovate and operate pervasive business models within a multitude of BMES with a multitude of security systems [20].

It is believed that the persuasive business model based on advanced security technology will indeed not only be important but add increasingly high value to the stakeholders involved and thereby give competitive advantage to businesses that act related to the new demands of persuasive business models. However we still have some challenge to overcome before we reach a final destination where all business models become persuasive—hereunder respecting security, trust, ethic challenges.

An even stronger focus on security, personal security, network based security technology, and business models that are continuously persuasive, in process and changing continuously in different BMES context sets business and researchers under high pressure to quickly find solutions—both technological and business wise to meet the needs of all kind of stakeholders for increasing agility, flexibility, individualization, privacy related to persuasiveness. A concept proposal for persuasive technology and business models which are independent of time, place, bodies and things—and at the same time are secure is proposed in Fig. 13.5. This is shown as proposed in a first generation proposal—a "ecosystem" of secure persuasive business model innovation showing the future context to come—the full integration between physical, digital and virtual persuasive business models operating in a secure context (Fig. 13.6).

An interesting synergy and spinoff of the above mentioned persuasive secure business model is a development to a completely new and changed understanding of persuasiveness, security and business models—taking us from a static, physical, digital and proactive security BM context—to a relational, interactive persuasive BM process perspective of individualized, integrated, automatic persuasive value proposition embedded with security delivery from BM's to users, customers, network, employees and things by all kinds of networks. Moving both the business model and security with time becomes suddenly imaginable and even possible in a world of 5G—an optimum for individuals and businesses valuing privacy, freedom, flexibility, agility and security at the same time. If so—it could really increase tremendously our global business potential and quality of life.

The fulfillment of embedding persuasive BM's anytime, anyplace, with anybody and any things seems to be achievable in the very near future. A last but not least important issue of the above mentioned models are security and business models related to the question where—which is not only a question of place as a

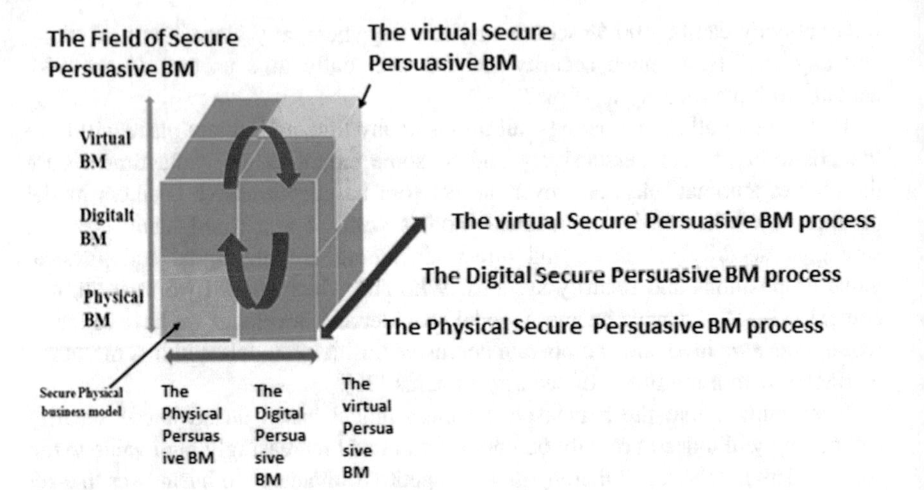

Fig. 13.6 A "ecosystem" of the secure persuasive physical, digital and virtual business model inspired by Whinston and Stall [21]

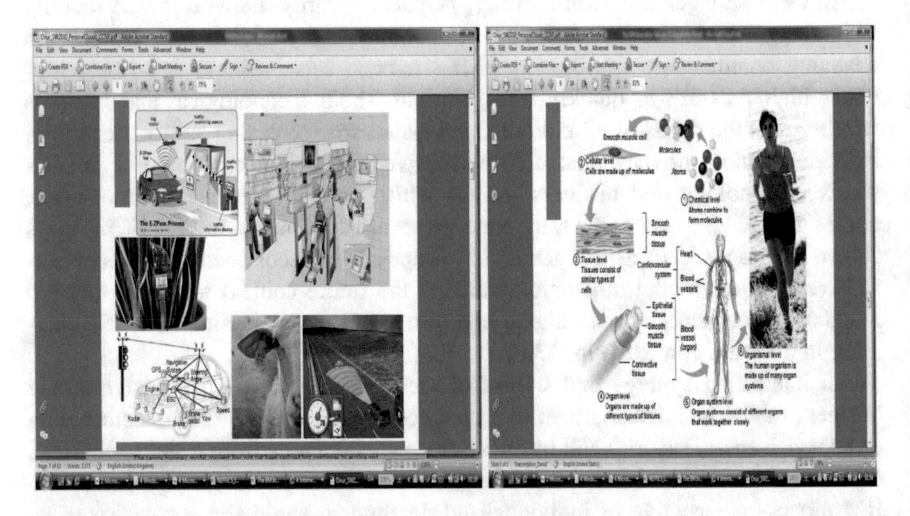

Fig. 13.7 Persuasive BM's outside and inside things and bodies any time, any place and with anything/anybody (Inspired from Onur [12] TU Delft presentation SW2010 [22]) and speakers from SW2015 [19]

geographical phenomenon. At SW 15 [19] it became even more visible than at the SW 10 [22] that future security and persuasive business models is not only a matter of security on the "surface" of things, places, people and time but also is about security and persuasive business models related to both inside and outside things, bodies and time (Fig. 13.7).

13.3 Conclusion

Persuasive business models we expect will become one of the most important BMI concepts for future business BMI and development. It will also be one of the most discussed topics related to ethics, trust a security—actually it has already begun (Google and EU, Yahoo, Facebook and China, Tobacco Industry and Global Health Care Society).

Persuasive Business Models are integrated with advanced persuasive business model technology—either product, service, production, process technology or all in combination. It is in this context that the fast evolvement of persuasive business models related to the vision of the secure, persuasive and sustainable business, business model ecosystem, business model ecosystems and world should be seen.

The secure persuasive business model concept has not yet been fully realized—we claim—but business and societies in general continue to evolve and embrace new perceptions, challenges and opportunities of the persuasive business model.

Secure persuasive business models can be physical, digital and virtual operating in a continuous process—integrated, connected and secure—delivering in a continuously process value propositions—where ever and whenever the user, customer, network partner, employees, things and business demands it. Persuasive business models operate typically together in a multi business model setup—a business model network collaboration outside and inside things and bodies in the future.

One must imagine that all 7 dimensions of a persuasive BM can and will change continually in the persuasive business model innovation processes related to the stakeholders demand, other persuasive business models and the strategy behind the persuasive business model. This makes it extremely difficult to measure the BM and the BMI process and thereby control and lead from outside either it is a user, customer, network partner, competitor or society.

Expectations are therefore that the majority of all future persuasive business models will be based on eventually a common understanding and vision of the secure world—secure persuasive business models based on secure persuasive technology and impact—anytime, anywhere with anything and anybody. This of course only if human being, global society and the business behind the business models including the technology will allow it.

Future security and persuasive business models is however not only a matter of security, ethics, trust and Business models on the "surface" of things, places, people and time but also is about these and business models related to both inside and outside things, bodies and time. How persuasive business models can be controlled in this context is therefore still for humans to perceive—see and sense. Both individual, across and together can change and form both their new business model contexture and their individual level of security, ethics and trust profile.

References

1. Chesbrough H (2006) Open business models. How to thrive in the new innovation landscape. Harvard Business School press
2. Fogg BJ (2003) Persuasive technology: using computers to change what we think and do. Morgan Kaufmann Publishers, San Francisco, CA, USA
3. Prasad R (2015) 5G. Springer
4. Fogg BJ, Eckles D (eds) (2007) Mobile persuasion: 20 perspectives on the future of behavior change. Stanford Captology Media, Stanford, California
5. Markides CC (2008) Game-changing strategies—how to create new market space in established industries by breaking the rules. Jossey-Bass Wiley, San Francisco
6. Afuah A, Tucci C (2003) Internet business models and strategies. McGraw Hill, Boston
7. Zott C, Amit R, Massa L (2011) The business model: recent developments and future research. "Electronic copy available at: http://ssrn.com/abstract=1674384"
8. Lindgren P, Søndergaard MK, Nelson M, Fogg BJ (2013) Persuasive business models. J Multi Bus Model Innovation Technol 1:70–98
9. Lindgren P, Taran Y, Boer H (2009) From single firm to network-based business model innovation. Int J Entrepreneurship Innovation Manage
10. Lindgren P, Taran Y, Saghaug KM 08-10-2010A Futuristic outlook on business models and business model innovation in a future green society
11. Lindgren P, Horn Rasmussen O (2013) The business model cube journal of multi business model innovation 3. Edition River Publisher Krebs, p 572
12. Onur E (2010) TU Delft presentation SW 2010 Presentation given by TU Delft at the Strategic workshop on sensor network at Florence July 2010. Inspired by ITU adapted from Nomura Research Institute
13. Chesbrough H, Rosenbloom RS (2000) The role of the business model in capturing value from innovation: evidence from XEROX corporation's technology spinoff companies. Ind Corp Change 11(3):529–555
14. Taran Y, Boer H, Lindgren P (2010) Managing risks in business model innovation processes. In: Presiding 11th international CINet conference practicing innovation in times of discontinuity Zürich, Switzerland, 5–7 September 2010
15. Fogg BJ (2012) Persuasive technology. Stanford University Press, Persuasive Technologies
16. Taran Y, Boer H, Lindgren P (2009) Theory building—towards an understanding of business model innovation processes. In: Proceedings of the international DRUID-DIME academy winter conference, economics and management of innovation, technology and organizational change
17. Abell DF (1980) Defining the business: the starting point of strategic planning. Prentice-Hall, Englewood Cliffs
18. Skarzynski P, Gibson R (2008) Innovation to the Core. Harvard Business School Publishing, Boston, MA
19. SW2015 Strategic Workshop 2015 held at Villa Mondragone Via Frascati, 51, 00040 Monte Porzio Catone, Napoli Italy. Human bond communications (HBC) Seventeenth strategic workshop (SW'15) 18–20 May 2015
20. Mucchi L (2010) Presentation physical layer cryptography and cognitive networks SW 2010 Florence—CNIT—University of Florence, Italy
21. Whinston AB, Stahl DO, Choi S (1997) The economics of electronic commerce. Macmillan Technical Publishing, Indianapolis, IN
22. SW2010 Strategic Workshop 2010 held at Hotel Relais Certosa, Florence, Italy, 26–28 May 2010–12. The strategic work shop 2010, Distributed and secure cloud clustering (DISC)

Chapter 14
Epilogue

Sudhir Dixit and Ramjee Prasad

The authors of the chapters of this book touched on a range of technology innovations to come, and how the wireless world might look like in the year 2050 and beyond. These concepts are discussed in multiple chapters. In this chapter, we briefly summarize those innovations in a more organized manner, and discuss their societal impacts. We briefly listed them in Chap. 1 in the atom diagram (Fig. 1.1).

All Applications over Internet

Internet will be pervasive and all applications (including multimedia, voice, and video) will be delivered over Internet to all types of user devices. The boundary between fixed and wireless access would be blurred and the device would connect over the most suitable and available access network based on its performance, cost and personal profile. Mobility will be taken for granted. It remains to be seen how the Internet would evolve in the future, such as towards the tactile Internet, 3-D internet, and where the content would be stored or cached to minimize latency. Applications in all industry verticals will proliferate to make life easy for all.

Ultra-broadband Mobile

While on the fixed access side, optical fiber would naturally continue to lead and play a major role to deliver gigabits and terabits of bandwidth, mobile broadband access will become ubiquitous delivering gigabits per second of bandwidth to each and every user, sufficient for most applications, including ultra-high-definition (UHD) media and its future generations. This would be made possible by massive MIMO and small and large cell technologies working in concert with each other.

S. Dixit (✉)
Skydoot, Inc., Woodside, CA, USA
e-mail: Sudhir.dixit@ieee.org

R. Prasad
Aalborg University, Aalborg, Denmark
e-mail: prasad@es.aau.dk

R. Prasad and S. Dixit (eds.), *Wireless World in 2050 and Beyond: A Window into the Future!*, Springer Series in Wireless Technology,
DOI 10.1007/978-3-319-42141-4_14

Context, Location and Personalization
The environment of the user and the thing, his/her/its location, and prior history of usage and interactions would play a major role on how and what services and applications are delivered that is customized. This would facilitate and make life easier.

Business Models
The prediction is that the business models would likely change completely in the next 30–40 years as we have seen this happen in the past decade, such as OTT services, advertising, on-line services. We expect that crowd and cloud would play an increasing role in a democratized Internet/web and will be part of the revenue sharing eco-system. This would and could happen in many ways.

Cloud
Public, private and hybrid clouds will be ubiquitous and will be in common use for storing content, running applications, and for compute and storage, for consumers, enterprises and things (e.g., sensors, machines). Network infrastructure elements, information technology and management, data centers, etc., would be abstracted as cloud instances. The users and enterprises would not need to invest in their own IT resources and management. Small and medium businesses will have undergone significant transformation due to cloud-based services and broadband access.

Big Data and Analytics
All types of data will be collected by the users, enterprises, application providers, public agencies, network providers, etc., and will be used to manage, deliver, and customize services with additional value and relevance by analyzing those data by powerful analytics engines. Some of this data and intelligence will be monetized and potentially shared by all the players in the value chain.

Sensors, IoE and Massive M2M Communications
Sensors and all kinds of things will be network connected and a vast variety of data from them will be collected to further mine the intelligence for various purposes. In addition, these devices and machines will be remotely diagnosed, updated, operated and managed without human intervention. Backend IT systems will play a major role to fully realize the potential of massive deployment of sensors in various forms, mobile devices, things and machines.

Massive MIMO
To meet the future requirements of spectrum and energy efficiency and massive capacity requirements to deliver wireless broadband to everyone will require massive MIMO for radio access. This would require significant innovation to control interference and physical design to accommodate in very small form factor equipment. The cloud-based processing and storage will make this possible to have just the radio heads and all signal processing to happen in the cloud. Developing such distributed and reconfigurable network architectures will be a challenge and one size (or solution) fits all scenarios would probably be not feasible.

Intelligent Convergent Infrastructure
The industry will continue to drive towards every networking element today (or the next generation Internet) to be implemented using the virtualization technology where the cloud, data centers and network elements will run on off-the-shelf (OTS) hardware. Most network and computing elements will be softwarized for flexibility, rapid provisioning, and rapid on-demand reconfiguration to meet the changing traffic and QoS demands. The network will extend up to wearables and potentially human bond communication, which is a holistic approach to describe and transmit the features of a subject in the way human perceives it through all five senses. The same intelligent infrastructure will be logically sliced end-to-end into several logical networks to meet specific requirements of the industry verticals. Network APIs will be used by the software layer to redesign the network layer.

Security, Privacy and Authentication
Security and privacy will continue to be issues, and problems with authentication will be solved through innovative solutions based on the fusion of bio-markers, i.e., biometrics, of users. Context will be an important determinant of the level of security and authentication. However, the perception of security and privacy as issues will undergo signification transformation and will be accepted as cost of convenience and being part of the connected world. Each successive generation will be more open to sharing their personal information and will deal with (and adopt to) the issues of security differently.

Peer-to-Peer (P2P) and Social
P2P networks, bypassing the operator, will increase in popularity. P2P and social applications will continue to gain popularity with the expectation that more innovative human-device (multi-modal) interaction technologies will be integrated with the applications to make them more intuitive and engaging. Search and recommendation engines will become more personalized and accurate leveraging the known network of users and service providers.

Energy Management
Global warming will lead to the awareness of ill- and unwanted impacts of ever-increasing consumption of energy on the environment and this awareness will be at par with the awareness to clean water and clean air today. All devices, including sensors, will use energy when required rather than being always on. This would be in addition to smart software implementations to minimize processor use and load distribution to minimize power consumption. In short, energy supply and demand will be managed smartly, where there are many initiatives already ongoing, such as smart grid, smart meter with Internet connectivity and analytics. A Code of Conduct would be in place for the citizens around the globe and public policy and regulations will be implemented to make devices, networks and applications as green as possible.

Finally, societal impact of communications and Internet technologies will be enormous. It is difficult to say how the countries, societies and cultures around the world will react and adopt to these changes, some positive and some negative. One thing is sure: the world will shrink with disappearing notion of distance and time, and it will be a more flat and democratized world with equal access to knowledge to all citizens! The challenge is upon us to adopt the good and not the unwanted negatives.

About the Book

Time and again, the best of the visionaries have been proven wrong about the innovations that have actually come to market and have impacted the human kind in unprecedented ways at an unimaginable pace. This book is an attempt to provide a long-range peek into the future of the emerging technology directions, end-to-end systems, long-term strategies, and the entrepreneurial and business models that will touch everything that we know of today, from humans to things to nature. This book is an attempt to predict and shape the future of wireless communications and enabling technologies in the 2050 and beyond time-frame by compiling thoughts and visions from the leading academics and scientists into a book. This is a "technology time capsule", being created in 2015, which the history would judge after 2050 as to how much we were (or not) on target. Thus, this is a visionary work by the scientific community to identify the interesting technologies to work on which could have unprecedented impact on the society and entire humankind.

R. Prasad and S. Dixit (eds.), *Wireless World in 2050 and Beyond: A Window into the Future!*, Springer Series in Wireless Technology,
DOI 10.1007/978-3-319-42141-4

Index

R. Prasad and S. Dixit (eds.), *Wireless World in 2050 and Beyond: A Window into the Future!*, Springer Series in Wireless Technology,
DOI 10.1007/978-3-319-42141-4

Printed in the United States
By Bookmasters